Synthesis Lectures on Human Language Technologies

Series Editor

Graeme Hirst, Department of Computer Science, University of Toronto, Toronto, ON, Canada

The series publishes topics relating to natural language processing, computational linguistics, information retrieval, and spoken language understanding. Emphasis is on important new techniques, on new applications, and on topics that combine two or more HLT subfields.

Federica Cavicchio

Emotion Detection in Natural Language Processing

 Springer

Federica Cavicchio
Department of Humanities
University of Salento
Lecce, Italy

ISSN 1947-4040   ISSN 1947-4059   (electronic)
Synthesis Lectures on Human Language Technologies
ISBN 978-3-031-72046-8   ISBN 978-3-031-72047-5   (eBook)
https://doi.org/10.1007/978-3-031-72047-5

This Springer imprint is published by the registered company Springer Nature Switzerland AG
The registered company address is: Gewerbestrasse 11, 6330 Cham, Switzerland

If disposing of this product, please recycle the paper.

Preface

Objectives and Scope of the Book

This book explores emotion detection within Natural Language Processing (NLP) by focusing on categorical and dimensional models of emotion detection. It provides a comprehensive overview of the resources, methods, and applications involved in emotion detection. While sentiment analysis offers a binary categorisation of emotions (positive/negative) and provides a computationally convenient method for understanding the overarching sentiment of a text, it is critical to acknowledge the instances where this approach may not suffice. Textual data involving complex human interactions demand a more nuanced understanding of emotions, considering contextual and linguistic subtleties. By focusing on specific emotion labels and their dimensions, this book seeks to provide a comprehensive picture of emotional expressions in textual data.

Traditionally, emotion theories such as the James-Lange theory (James, 1884; Lange, 1887) and the Cannon-Bard theory (Cannon, 1927; Bard, 1928) have primarily approached emotions from a physiological standpoint. Ekman and Friesen (1971), influenced by Darwin (1872), posited that emotions, particularly the so-called "basic emotions", such as happiness and anger, are biologically ingrained and defined by identifiable neural and expressive patterns (e.g., facial expressions).

In contrast, social psychology theories have emphasised the role of cognitive appraisal in decoding and labelling emotional states (Smith & Lazarus, 1993). Embodied cognition theories of emotion (e.g., Barsalou, 2008) posit emotions as mentally constructed states, asserting that recognising an emotion relies on the conceptual knowledge acquired from past experiences (Barrett et al., 2007). Thus, diving into emotion studies reveals an ongoing debate on how we recognise emotions: categorically (basic/discrete emotions) or dimensionally (e.g., the valence, activation, and dominance dimensions of the emotion experience).

Categorical theories of emotion, supported by Izard (2007), focus on a restricted set of innate, hard-wired basic emotions, and prescribe rigid affect programmes leading to prototypical response patterns. Conversely, dimensional and appraisal theories, advocated by researchers such as Scherer (2001) and Frijda (1986), lean towards a dynamic architecture based on appraisal and motivational changes that affect both the autonomic and somatic nervous systems. This approach involves an automatic, unconscious appraisal process rooted in previous experiences and long-term memory; it considers emotions as dynamic representations that can be assigned to emotional categories or labelled with various linguistic expressions.

Amidst these theoretical discussions, there is a widespread acknowledgement that valence—categorising emotions as positive or negative—is a fundamental facet of the emotional experience (Russell, 2003). Notwithstanding its crucial role across appraisal theories, the concept of valence has faced criticism for its binary nature. While some suggest that the concept of valence might be entwined with cultural and ethical norms (Solomon & Stone, 2002), other authors (e.g., Widen & Russell, 2003) posit a hierarchical division of emotional concept knowledge into different tiers: overarching (valence) and specific (emotional labels).

Mohammad (2020) and Sailunez et al. (2018) have explored the theoretical aspects and the NLP applications for sentiment analysis and categorical emotion detection, respectively. Mohammad (2020) focused primarily on sentiment analysis, discussing the role of sentiment analysis in areas such as public health and social science. He also addresses current challenges such as the lack of large, labelled datasets and how to detect the subjective nature of sentiment. On the other hand, Sailunez et al. (2018) provided a thorough survey of emotion detection from both text and speech, reviewing various methods and models for identifying emotion categories. Their work emphasises the practical applications of emotion detection in e-learning, mental health, and customer service sectors.

This book touches on similar themes to those in Mohammad's (2020) and Sailunez et al.'s (2018) works. However, in this book, we go further by engaging in the theoretical debate between categorical and dimensional emotion models and their application in NLP. Our theoretical exploration provides insights not thoroughly covered in previous studies. In particular, this book highlights the importance of exploring a range of emotion models beyond the positive-negative and discrete categories, aiming to provide a more detailed overview of how psychological models of emotion expressions can be applied in NLP.

Emotions, by nature, encapsulate a rich, multifaceted spectrum of human experience that cannot always be accurately or meaningfully reduced to a positive/negative dichotomy or a single label. Numerous scenarios, particularly those involving complex human interactions and communication, demand a more nuanced understanding and processing of emotions. For example, customer feedback, hate speech detection, or literary analyses often encompass a blend of emotions that a binary positive/negative classification might overlook. Furthermore, contextual, and linguistic subtleties can influence the binary and categorical emotional expressions and their perception. Therefore, we necessitate a more detailed, multilayered approach to emotion analysis. Hence, while sentiment analysis serves as a valuable starting point, specific contexts and complex applications call for a more intricate, exhaustive, and multidimensional approach to emotion detection.

Emotions weave a complex tapestry deeply intertwined with our communicative expressions and, consequently, guide the meaning embedded within language. By focusing on specific emotions such as joy, anger, or surprise, we can directly label feelings expressed in words, making them easier to understand and discuss. On the other hand, looking at emotion dimensions means we will explore how positive or negative, moving or calming, empowering or powerless those emotions are, which can help catch the subtle shades of feelings that might not be clear at first glance. By using both approaches, categorical and dimensional, we will get a complete picture, and will identify clear-cut emotions while, at the same time, we will uncover the subtle layers of feelings hidden in the text.

In the next paragraphs, we introduce the key concepts and terminology of the expression of emotions. We will then briefly describe emotion detection systems and their applications.

Lecce, Italy   Federica Cavicchio

References

Bard, P. (1928). A diencephalic mechanism for the expression of rage with special reference to the sympathetic nervous system. *American Journal of Physiology, 84*(3), 490–516.

Barrett, L. F., Lindquist, K. A., Bliss-Moreau, E., Duncan, S., Gendron, M., Mize, J., & Brennan, L. (2007). Of Mice and Men: Natural kinds of emotions in the mammalian brain? A response to Panksepp & Izard. *Perspectives on Psychological Science, 2*, 297–312.

Barsalou, L. W. (2008). Grounded cognition. *Annual Review of Psychology, 59*, 617–645.

Cannon, W. B. (1927). The James-Lange theory of emotions: A critical examination and an alternative theory. *The American Journal of Psychology, 39*(1/4), 106–124.

Darwin, C. (1872). *The expression of emotions in man and animals.* John Murray.

Ekman, P., & Friesen, W. V. (1971). Constants across cultures in the face and emotion. *Journal of Personality and Social Psychology, 17*, 124–129.

Frijda, N. H. (1986). *The emotions.* Cambridge University Press.

Izard, C. E. (2007). Basic emotions, natural kinds, emotion schemas, and a new paradigm. *Perspectives on Psychological Science, 2*, 260–280.

James, W. (1884). What is an Emotion? Mind, 9, 188–205.

Lange, C. G. (1887). *Ueber Gemüthsbewgungen. Eine Psycho-Physiologische Studie.* Theodor Thomas.

Mohammad, S. (2020). Sentiment analysis: Automatically detecting valence, emotions, and other affectual states from text. *Emotion Measurement* (2nd Edition), 323–379.

Russell, J. A. (2003). Core affect and the psychological construction of emotion. *Psychological Review, 110*(1), 145–172.

Sailunaz, K., Dhaliwal, M., Rokne, J., & Alhajj, R. (2018). Emotion detec-tion from text and speech: A survey. *Social Network Analysis and Mining, 8*, 28.

Scherer, K. R. (2001). Appraisal considered as a process of multilevel sequential checking. In K. R. Scherer, A. Schorr, & T. Johnstone (Eds.), *Appraisal Processes in Emotion: Theory, Methods, Research*, 94, Oxford University Press.

Smith, C. A., & Lazarus, R. S. (1993). Appraisal components, core relational themes, and the emotions. *Cognition & Emotion, 7*(3–4), 233–269.

Solomon, R., & Stone, L. (2002). On "positive" and "negative" emotions. *Journal for the Theory of Social Behaviour, 32*, 417–435.

Widen, S. C., & Russell, J. A. (2003). A closer look at preschoolers' freely produced labels for facial expressions. *Developmental Psychology, 39*(1), 114–128.

Contents

Introduction

1.1 Emotion Concepts and Terminology

Tyng et al. (2017) organized these terms in relation to their relevance for emerging technologies, helping to clarify how emotions can be detected and classified by machines. Understanding these terms can also help us get a clearer picture of the complexity of the human emotions we want to detect. The rise of emotion-detecting social robots and intelligent conversational agents highlight the interdisciplinary nature of emotion detection, integrating insights from psychology, sociology, human–computer interaction, and data mining: chatbots can now tailor responses based on the customer's emotions; in healthcare, patient expressions can provide vital health clues; and in public policy, sentiment analysis can inform governance decisions.

The term *sentiment* usually refers to self-focused emotions such as sadness, ecstasy, and nostalgia and can also denote opinions strongly influenced by emotions.

The term *emotion* refers to a mix of observable and personal factors that are controlled by our autonomous system, our cognition, and our body. Emotions help shape the way we think about and react to the world. Emotions also guide how we communicate, our goal-oriented actions, and how we adjust to new situations, though they can be unpredictable. For example, happiness is a complex psychophysiological experience that involves an initial appraisal of a situation as beneficial or fulfilling our goals. This cognitive appraisal leads to a happiness response that includes a synchronized response pattern across multiple systems. Happiness elicits changes in the autonomic nervous system consistent with the positive valence of the emotion. These changes in the autonomic nervous system could result in a decrease in stress-related hormones and a reduction in heart rate variability. As regards facial expressions, there is the activation of the muscles involved in smiling, such as the zygomatic major, which pulls the lips upward, and the orbicularis oculi, which creates crow's feet around the eyes—what Ekman and Friesen (1978) called the Duchenne

F. Cavicchio, *Emotion Detection in Natural Language Processing*, Synthesis Lectures on Human Language Technologies, https://doi.org/10.1007/978-3-031-72047-5_1

smile, a genuine indicator of happiness. Behaviourally, happiness can lead to increased social behaviour, an inclination toward sharing with others, and a greater likelihood of engaging in creative or exploratory activities. All these responses—expressive, physiological, and behavioural—are interrelated and influence each other, contributing to what we collectively label happiness.

Feelings generally refer to the subjective experience of emotions. Feelings are the conscious perception of emotional responses. While emotions can trigger physical reactions and are often observable through behaviour, feelings are how we interpret and label these internal experiences. For example, the emotion of fear may lead to a feeling of dread or anxiety, depending on the person and the context. Feelings are shaped by a person's past experiences, beliefs, memories, and thoughts associated with that emotion. They are the mental portrayal of the observable or physiological changes in the body that accompany an emotion and are often intertwined with our thoughts and mood. Since feelings are a mental representation of the body's physical responses to emotions and the cognitive assessment of the situation, they are not as observable as emotions and physical responses. However, they are intensely personal and can vary widely from one individual to another, even in response to the same event. Feelings are influenced by individual personality, past experiences, and the meaning that the event holds for the person.

Moods are more diffuse and less intense than emotions and can last hours, days, or longer. A single specific event does not typically trigger moods. Moods are usually more generalized, often influencing how we perceive the world. They can influence our overall mental state and can predispose us to experiencing certain emotions instead of others. For example, when we are in a bad mood, we may be more likely to feel irritable or frustrated during the day.

Affect is considered the basic building block of more complex emotional states (emotions, feelings and moods). Affect has two key components: valence and arousal. Valence refers to the intrinsic attractiveness (positive valence) or averseness (negative valence) of an event, situation or object. The valence of the affective response can influence a person's willingness to approach or avoid a situation or object. Arousal indicates the physiological state elicited from the situation or object. In psychological research, emotions are plotted along the dimensions of valence and arousal to create a more nuanced understanding of affective states. Affect is usually studied through its influence on cognition and behaviour.

In the following we summarise the conceptual frameworks and terminology associated with human emotions and includes examples to illustrate each category:

Sentiment: It refers to self-focused emotions such as sadness, joy, or nostalgia; it also refers to opinions influenced by emotions. For example, a nostalgic sentiment may involve a longing for past experiences or childhood memories.

- *I miss the Nineties when everything was simpler.*

Emotion: It is a mix of observable and internal factors influenced by autonomous systems, cognition, and the body. It involves cognitive appraisals and physiological responses. For instance, anger involves increased heart rate and tense muscles, often accompanied by aggressive behavior.

- *Her heart raced, and her jaw tensed as she clenched her fists in anger.*

Feelings: It is the subjective experience of emotions, representing the conscious perception of emotional responses. For example, love may lead to feelings of warmth, affection, and attachment.

- *Feeling a deep sense of warmth and connection, he knew he was falling in love.*

Moods: It refers to longer lasting and less intense states than emotions, influencing overall mental state and predisposition to emotions. For instance, being in a gloomy mood may make one more susceptible to feeling sad.

- *He couldn't shake off the gloominess that hung over him all day.*

Affects: It is the basic building block of emotions, feelings, and moods, characterized by valence (positive/negative) and arousal (physiological state). For example, high arousal and positive valence correspond to excitement, while high arousal and negative valence correspond to anxiety.

- *His anxiety spiked, leaving him restless and uneasy all night.*

1.2   Emotion Detection in NLP

Emotion detection in NLP is a particularly challenging task due to the dual nature of emotion expressions. Emotion communication comprises the personal experience of emotion and how it is communicated. As such, when individuals express emotions, they often convey their internal emotional state to express a specific intent or to evoke a particular response from their interlocutors.

The communicative intent can significantly alter how emotions are presented, usually adding layers of nuance. For instance, a short message such as *I'm fine* can carry a wide range of emotional undertones depending on the communication context, tone of the conversation/interaction, and the relationship between the interactors, which NLP models may struggle to interpret accurately. Generally, in shorter texts, such as tweets or text messages, the brevity of the messages limits the contextual clues and emotional cues available, making it difficult to distinguish an expression of emotion.

Categorical and dimensional emotion models provide varied frameworks for interpreting emotions. Over the years, researchers have explored numerous NLP methods, from lexicon-based approaches to machine learning, to improve the contextual understanding of emotions.

A first, significant advancements in emotion analysis started with the creation of ISEAR (Scherer & Wallbott, 1994), which marked the introduction of the first emotion dataset featuring sentences labelled with a range of emotional states. This initiative provided a solid foundation for subsequent developments in emotion detection within texts. Valitutti et al. (2004) expanded the toolkit available for researchers by presenting WordNet-AFFECT, a knowledge-based network of concepts comprising affective words and emotion labels. The NRC Lexicon (Mohammad & Turney, 2013) encompasses a collection of 14,000 words, each assigned to a specific emotion and sentiment, classifying words according to their emotional state. The NRX Lexicon is effective at identifying emotions but could not detect emotion intensity. Thus, the NRC-valence arousal-dominance lexicon was created (Mohammad, 2018). This lexicon can attribute emotions with their respective intensities by integrating emotions' dimensions (valence, arousal, and dominance). This progression in lexical resources paved the way for developing emotion detection systems.

Emotion detection through rule-based methods depends on predefined rules and patterns, often established with the help of lexicons and bodies of text known as corpora. The keyword-based method (Ma et al., 2005; Tao, 2004) initiated the trend of rule-based approaches to emotion detection in text, mapping keywords to emotions by drawing from resources such as WordNet-AFFECT. This method involved preprocessing texts to align them with dataset annotations and analysing emotional keyword intensity and negations. In parallel, the rule-based technique used linguistic rules—crafted after text cleansing and Part of Speech tagging—to identify emotions with a probabilistic approach, as demonstrated by the works of Perikos and Hatzilygeroudis (2013) and Shivhare et al. (2015).

Aman and Szpakowicz (2008) made significant contributions to the early development of emotion detection in texts. They explored the task of automatically categorising sentences into Ekman's six basic emotion categories. Their experiments utilised corpus-based features and features derived from two emotion lexicons. One lexicon used Roget's Thesaurus (1852), while the other comprised words extracted from WordNet-AFFECT (Valitutti et al., 2004). Experiments on data obtained from blogs revealed that combining corpus-based features with emotion-related features resulted in superior classification performance for all the emotions.

Building on these foundations, Ghazi et al. (2010) explored the task of automatic classification of texts. Their approach hierarchically arranged neutrality, polarity, and emotions, which they tested on two datasets. This hierarchical approach outperformed the corresponding non-hierarchical, flat approach. Further expanding on their work, Ghazi et al. (2012) used a set of words labelled with their corresponding emotions as a basis for automatic emotion detection in sentences. They also recognised that context must be considered, as words are interrelated and mutually influence their affect-related interpretation. Ghazi et al.'s method accounts for a word's contextual emotion and the sentence's syntactic structure to classify sentences by emotion classes.

However, there are challenges with rule-based emotion recognition, such as limited emotion labels in lexicons, ambiguity of keywords, and lack of consideration for the linguistic context. Word embeddings, such as Word2Vec proposed by Mikolov et al. (2013), represent a significant advancement in context disambiguation by using neural networks to capture the context of a specific word in a text. Word embeddings depart from traditional rule-based systems, which struggle with context and the subtleties of language, such as polysemy and synonymy. Word2Vec and similar technologies capture semantic and syntactic word relationships based on the word context. Shifting the focus on context leads to an improved text analysis. The advent of transformer architecture (Vaswani et al., 2017) led to the development of pre-trained models such as BERT (Devlin et al., 2019). These models dramatically advanced text classification and analysis, utilizing the deep learning-based approach to process labelled and unlabelled data through intricate neural network layers. The machine learning-based approach showed the potential for systems to autonomously refine their capabilities through experience, as seen in the research by Aman and Szpakowicz (2007) and Ghazi et al. (2010). This method hinges on extracting salient text features and training algorithms with emotion labels to predict emotional states in new data. Deep learning, a subset of this approach, advanced these capabilities even further, with researchers such as De Bruyne et al. (2018) and Suhasini and Srinivasu (2020) employing complex multi-layered neural networks for even more nuanced and accurate emotion detection.

Emotions, inherently complex and often fuzzy, challenge the boundaries of these techniques. Emotions' subjective nature varies widely across individuals and contexts, necessitating algorithms that can navigate this variability. Detecting emotions remains challenging, highlighting the need for ongoing improvements in algorithmic sensitivity and adaptability. The culmination of these developments represents the current state-of-the-art in NLP for emotion analysis and detection, where deep learning architectures and transfer learning techniques strife to deliver more precise and context-aware emotion detection.

1.3 Applications

The methodologies described so far help us to understand emotion detection and expand its applicability across various sectors. In fact, emotion detection plays a crucial role in management and marketing, user engagement, healthcare and education. These emotion detection techniques enable the analysis of vast quantity of text data in systems that manage product and service reviews, social media interactions, and in conversational chatbots. We begin with the latter.

1.3.1 Conversational Interfaces/Chatbots

Conversational interfaces, also known as chatbots, dialogue systems, or intelligent virtual agents, among other terms, are HCI models equipped with AI. Early models of chatbots, such as ALICE and ELIZA, employed pattern matching with logical clauses to guide conversations. More recently, machine learning models leverage data to predict conversational actions without needing domain-specific rules, using Bayesian networks, neural networks, and Markov models. As detailed by Adikari et al. (2019), chatbots can employ word embedding, Markov chains, and NLP to discern emotions. Chatbots are designed to engage in conversations with users using natural language, either spoken or written, and are currently deployed in various sectors, including e-commerce, education, and health-care. They are employed for their availability, scalability, and cost efficiency, which boosts competitive advantage. These systems also elevate customer service by providing friendly, flexible, and effective assistance, aiming to generate more interactive responses. With the evolution of machine learning, chatbots have evolved to respond empathetically to users, opening new avenues. In education, chatbots can act as online learning tutors, delivering instructional content (Song et al., 2017), course applications, schedules, and academic information (Hien et al., 2018). In healthcare, conversational agents can convey medi-cal information or interpret clinical results (Montenegro et al., 2019), aiding clinicians in symptom identification and diagnostic processes (Pacheco-Lorenzo et al., 2021). Conver-sational assistants can also serve purposes from diet management to mental health support (Fadhil et al., 2019).

Understanding emotions is crucial for Chatbots as it allows them to tailor their responses to the user's emotional state, leading to meaningful and satisfying interactions. Emotion detection enables Chatbots to respond empathetically, reducing user frustration, building trust, and improving the user experience. For example, in the healthcare and mental health sectors, recognising emotions can significantly enhance the quality of the virtual assistants' support. It is also worth noticing that chatbots are increasingly used for mental health support by users, and more and more studies evaluate their effectiveness in helping with mental health crises (for example, Fitzpatrick et al., 2017; Inkster et al., 2018).

To achieve human-like interaction, Chatbots integrate Natural Language Understanding (NLU), Dialogue Management (DM), and Natural Language Generation (NLG). The NLU interprets the user's intent, DM organizes the flow of conversation, considering context and domain knowledge, and NLG crafts the textual responses. However, chatbots are still challenged in interpreting user intent, which disrupts the conversation flow and compli-cated intent classification, which can lead to user frustration. Finally, while e-commerce has seen extensive chatbot integration, other sectors, such as healthcare, finance, and education, are yet to capitalize on this technology fully.

1.3.2 Social Media

Social networks such as Facebook, Instagram, X-Twitter, Reddit, YouTube, and WhatsApp dominate the Western market of online communication. Instagram, Reddit, and X-Twitter, specifically, are extensively utilized for microblogging. Online social media has emerged as the channels for the dissemination and exchange of personal and global information and ideas through posts that encompass text, visuals, audio, and video. Emotion Detection can help unravel the complex interplay of emotional undertones of shared sentiments and viewpoints. Such analysis could shed light on the effects of social media on social tendencies and individual behaviours. Numerous fields, such as business intelligence and behavioural analysis, utilize social media for diverse applications, including sentiment analysis, epidemic and crime detection.

Pool and Nissim (2016) investigated the classification of emotions via Facebook reactions within a remotely supervised setup, utilizing a Support Vector Machine (SVM) to distinguish emotions such as anger and surprise. Yasmina et al. (2016) tackled the intricacies of conversational style and linguistic evolution in chat environments by pinpointing emotions from YouTube comments. Sailunaz and Alhajj (2019) devised a recommendation system through sentiment analysis and emotion detection in X-Twitter feeds. The system discerns various emotions, such as anger and joy, and sentiments' polarity. Additionally, the system tailors its recommendations to the users by assessing their use sway and considering a combination of the users' and their tweets' characteristics. Wang et al. (2017) explored the bonds among individuals discussing a product or service, introducing the concept of *sentiment community*. Gupta et al. (2021) examined the public's sentiment during epidemics, assessing the impact of weather on COVID-19's spread based on X-Twitter data. Their topic modelling also revealed significant pandemic-related discussion themes. Sun et al. (2015) proposed a social regulation strategy to enhance recommendation systems by incorporating user relationship data to determine trusted network friends.

Nevertheless, the detection of emotions in social media emotion still faces numerous challenges. The informal, orthographic, and grammatical error-prone language on these platforms poses a significant hurdle, along with the subjectivity and ambiguity of the emotion expressions. The complexity of emotions, requiring human interpretation, makes the crafting of a comprehensive dataset for training these systems a daunting task, mainly due to the vast range of topics and contexts in social media data.

1.3.3 Review Systems

Business Intelligence (see Correira Loureiro et al., 2021) describes a variety of methodologies, frameworks, and approaches aimed at converting data into meaningful insights that enable businesses to operate with greater efficiency. It comprises a collection of applications and services that translate data into practical knowledge and insights. The integration of business intelligence with social media data offers significant advantages. Insights derived from social media, such as reviews on products and services, tweets, and various forms of engagement, can guide companies in delivering superior quality products and services. Early-stage feedback on product design and development is particularly invaluable for incorporating customer insights into the conceptual stages of product creation. User experience pertains to a person's reactions and feelings towards using a particular product, system, or service, which is increasingly acknowledged as a competitive edge in product and service design. Since customer satisfaction is inversely related to customer churn, businesses heavily invest in analysing and enhancing customer satisfaction.

While the review systems domain is not exclusively tied to emotion detection, it does address user sentiments such as neutrality, positivity, or negativity, which are indicative of underlying feelings. Thus, there is a recognized need to bridge the gap between sentiments and specific emotions by leveraging linguistic elements in feedback and reviews, categorizing them into distinct emotional categories. Emotion and sentiment detection from reviews has been applied in various domains, including movie, hotel, and e-commerce product reviews. It is interesting to note that, so far, emotion detection has been investigated more in the field of product reviews than in service reviews. Bai et al. (2019) specifically investigated how insights from online customer reviews can inform and improve product design. They developed a multi-faceted framework to identify key elements of user feedback for product development and proposed methods to build a knowledge base from these reviews. Their approach comprises three stages: discovery of user feedback from individual reviews, aggregation with similar data, and the formalization of a network that draws causal links between different sets of user feedback data. Mirtalaie et al. (2018) aimed to automate the analysis of customer reviews to extract and measure sentiments about specific product features, resulting in the Pros/Cons Sentiment Analyzer framework that relies on dependency relations to extract sentiment information from reviews.

1.3.4 Analysis of Literary Texts

While most research has traditionally concentrated on evaluating product reviews, tweets, and emails, more and more studies are exploring emotion expressions in literary texts (e.g. Hsu et al., 2015; Mohammad, 2012; Whissell, 2003). These studies commonly treat the analysis of emotions in literature as a classification task, aiming to label a text or a text segment with an emotion term or the emotion valence (positive/negative). These studies primarily focused on identifying distinct emotions such as love, anger, and fear and predominantly involved classifying texts or text segments into these categories. For example, Mohammad (2012) analyzed the polarity and density of emotional words in novels and fairy tales, using the NRC Emotion Lexicon to tag words with polarity and distinct emotions like joy and sadness. The study employed an emotion analysis tool to track the frequency of words associated with specific emotions and compare how these emotional expressions vary across different literary genres. Despite concentrating on distinct emotions through specific emotion-laden words, this approach provided a limited insights into literary critic, falling short in effectively modelling how emotions develop through a narrative. More recently, the studies on emotion expression in literature have investigated the emotional tones expressed throughout a story, trying to capture the emotions expressed in narrative arcs, the characters development, and the evolving emotional trajectory of the narratives. Reagan et al. (2016) examined the emotional arcs of narrative fiction using a sliding window of sentences. They found six fundamental narrative arcs through 1327 books from Project Gutenberg: Rags to Riches (rise), tragedy (or Riches to Rags, fall), Man in a Hole (fall-rise), Icarus (rise-fall), Cinderella (rise-fall-rise) and Oedipus (fall-rise-fall).

The aforementioned studies share the approach to detecting emotion terms or emotion valence in literature as a classification problem. Their objective is to label a text or a segment with an emotion or an emotion valence label, which can provide insights into the author's perspective or offer quantitative data for literary analysis. In contrast, in their study Watson et al. (2020) seek to model the emotion experience throughout the text, aiming to predict a specific numerical value (indicating the level of positive or negative emotion) for each sentence in the narrative sequence. Watson et al. examined a number of different context sentences to identify the most informative range that could predict the emotional valence of the target sentence. Not surprisingly, they found that the sentences closer to the target sentence they wanted to label were the most informative in detecting the target sentence's emotional valence. Similarly, Christ et al. (2022) traced the valence/arousal trajectories throughout different stories and, integrating sentence context, found that the sentences that precede the target sentence proved generally more valuable than the sentences that followed in capturing the emotional tone of the target sentence.

1.3.5   Mental Health Support

As we have seen above, identifying emotions within their specific contexts is a complex endeavour, primarily because of the long-term dependencies occurring in textual sequences from which we can infer meaning. These dependencies necessitate a detailed analysis of how emotions unfold and interact throughout client-therapist conversations or self-report stories.

Accurate mapping of emotional progressions in the text is essential for developing robust natural language processing models that reliably detect subtle signs of mental health issues. Research in online mental health has predominantly focused on depression (Tadesse et al., 2019) and, more recently, eating disorders (Huisman et al., 2024). Typically, these studies analyse transcripts from interactions between clients and mental health professionals, aiming to classify and categorize emotions according to established psychological theories such as cognitive behavioural therapy (Provoost et al., 2019).

Historically, detecting mental illness has relied on patients seeking psychological support proactively. However, the lack of developed mental health services in remote and underdeveloped areas, along with the high costs of diagnostics, often prevents timely access to treatment. With the growth of social media, an increasing number of people are using platforms such as Reddit, which features many groups of self-identifying individuals with depression and anxiety. Such online presence provides a substantial dataset of negative emotional expressions. As a result, the automated recognition and diagnosis of depression by analysing emotional cues has become increasingly important. Furthermore, contemporary research in mental health often uses natural language processing techniques to analyse social media content. Skaik et al. (2021) discuss methods for data collection, classification systems, and metrics for evaluating mental health in social media posts. Rissola et al. (2021) explore computational methods for assessing mental states online. Researchers also have specifically addressed the detection of suicide risk (e.g. Castillo-Sanchez et al., 2020) and depression (e. g. Giuntini et al., 2020). Finally, recent studies such as Sawhney et al. (2020), have focused on detecting significant emotion term trends over time as a way to identify mental suicide ideation on social media.

1.3.6   Humour, Sarcasm and Hate Speech

In the previous paragraphs, we have seen that emotion detection can improve recommendation systems and customer relationship management by revealing customers' likes and dislikes or eliminating the item recommendations that get negative feedback from the customers. However, most social content on the Web consists of sarcasm and irony. In social media, various people usually employ irony or sarcasm to communicate their emotions, making it difficult to analyse their real intent. Humour detection has emerged as a captivating and challenging research area in natural language processing. Humour, a distinctive and complex aspect of human communication, not only serves as a means of entertainment but also plays a crucial role in facilitating social interactions and conveying information in daily communication.

Since humour is a complex and content-dependent aspect of language use, which often involves puns, wordplay, irony, sarcasm, satire, incongruity, and cultural interpretations, the intricate nature of humour often poses challenges for humour detection using deep learning. Linguistic mechanisms and devices are fundamental to the creation and comprehension of humour. Various linguistic tools are employed to produce humorous effects, including puns, wordplay, irony, sarcasm, satire, incongruity, and parody (Martin & Ford, 2018). Puns, for instance, exploit words' multiple meanings or sounds to create humorous ambiguity. In addition, the role of linguistic context in humour comprehension must be considered. Humour depends on contextual cues and shared knowledge, enabling the listener to interpret and appreciate the intended humour. Contextual information, such as the setting, participants, and preceding discourse, provides crucial cues for decoding the humorous elements in a conversation or text. Understanding the social, cultural, and linguistic context is essential for recognising the incongruity or deviation from expectations that often underlies humorous situations (Clark, 1996). Additionally, pragmatic principles, including implicature and presupposition, play a significant role in humour interpretation. Conversational implicatures, based on Grice's Cooperative Principle, involve inferring meanings that go beyond the literal content of the utterance, contributing to the humorous effect.

Beyond linguistic context, the broader cultural and situational context also influences the production and reception of humour. Cultural references and shared knowledge serve as rich sources of humour. Jokes and humorous expressions often draw upon culturally specific knowledge, historical events, or stereotypes that resonate with a particular audience. Since many factors are linked to humour, one avenue to detect irony is emotional content, such as emoticons and writing styles, punctuation, capitalisation, and emotional words (Mahajan & Zaveri, 2020).

Similarly challenging is capturing sarcasm in texts. A sarcastic statement represents a conflict between the individual's communicative intent and sentence composition. For example, the sarcastic expression, *I love working on holidays!* shows a conflict between the explicit statement *on holidays* and the expression *work*. The contradiction and the sentiment polarity shift prove that sarcasm is a unique case of communication. Sarcasm is extraordinarily contextual and topic-reliant, and as a result, some contextual clues and shifts in emotion polarity can assist in sarcasm identification in the text. Insufficient knowledge of the context and the specific topic will result in difficulty detecting sarcastic utterances. To overcome the limitations related to context, Agrawal et al. (2020) incorporated emotion transition detection (e.g. from a positive to a negative emotion or from happy to angry) to accurately identify sarcastic text.

Another challenging task is the detection of hate speech. Hate speech, defined as derogatory communication targeting individuals or groups based on race, ethnicity, gender, sexual orientation, nationality, or religion, has been a significant concern since the rise of the internet and social media. Platforms like X-Twitter and Facebook, which allow users to share thoughts freely and often anonymously, foster vibrant discussions but can also facilitate harmful and offensive expressions. The sheer volume of daily content on these platforms makes manual monitoring and filtering such speech difficult. Hate speech manifests in both implicit and explicit forms. For instance, the statement *All* [members of a specific group of people] *are criminals and should be banned from our country* is an example of explicit hate speech because it directly promotes stereotypes and calls for discrimination against a particular ethnic group. Conversely, the statement *Certain neighborhoods are just not safe anymore thanks to the quality of people moving in* employs coded language to subtly suggest negative stereotypes about immigrants or ethnic groups, exemplifying implicit hate speech. While detecting explicit hate speech is relatively straightforward due to its directness, identifying implicit hate speech is more challenging, requiring a deep analysis of the overall context and linguistic nuances.

Traditional detection methods have primarily utilised lexicon-based and corpus-based features, relying on the specific characteristics and co-occurrence of terms within a domain to identify hateful content. However, recent research has expanded to include emotional features, although these efforts have been generally limited to specific emotional categories and singular "hate" categories (Martins et al., 2018; Safi Samghabadi et al., 2020). Detecting implicit hate speech involves data-driven approaches that leverage language models to discern semantic features (El Sherief et al. 2021) and basic and fine-grained emotion terms (Jafari et al., 2023). These approaches show that incorporating emotional nuances and sentiment polarity can significantly enhance the identification and understanding of hidden patterns in hate speech, offering insights into its subtle manifestations and implications.

In the following we present applications utilising NLP for emotion detection. We included an example sentence for each application to illustrate the emotions that the NLP systems are designed to detect:

Conversational interfaces/Chatbots: they are AI-based models used in e-commerce, healthcare, and education. They employ NLP for understanding and responding to the user's emotions:
- *I need help with my prescription.*
- *I'm feeling anxious about my doctor's appointment today.*

Social media: emotions in social media posts can be used in behavioral analysis and trend monitoring:
- *Struggling to keep up with the fitness challenge, but not giving up!*
- *I'm furious about the statements made by some of the candidates at the debate*

Review Systems: they analyse customer reviews for insights on product and service design. Emotion detection in reviews helps in understanding customer satisfaction and product feedback:
- *This product exceeded my expectations!*
- *The in-shop customer service was terrible, I'm very disappointed.*

Literary Studies: applying NLP techniques we can analyse emotion expression in literature. For example, we can study the development of themes and characters across narrative arcs:
- *The protagonist's journey through grief and redemption was deeply moving.*
- *The story's dark humor kept me hooked.*

Mental Health: using emotion analysis we can detect and monitor mental health conditions. NLP models can identify through emotion cues mental health changes, which could help in diagnosing depression and anxiety, and potentially alerting professionals to the emergence of mental crises:
- *I'm feeling hopeless and don't see a way out.*
- *Today was a good day, I felt more positive.*

Humor, Sarcasm, and Hate Speech Detection: NLP applications to identify humor, sarcasm, and hate speech by detecting emotions and emotion polarity changes. These tools are essential for content moderation on social platforms. They help disambiguate the text and distinguish harmful speech from playful banter, contributing to a safer online environment while preserving free speech:

- **Sarcasm:** *Sure, because everyone loves working late on a Friday...*
- **Hate Speech:** *Those people are ruining everything and we should silence them.*

In the following chapters, we provide a comprehensive methodological perspective on emotion detection, critically examining the rapid advancement of AI methods and the innovations aimed at accurately capturing emotional expressions in language. We also address the challenges posed by linguistic subtleties and emotion granularity, and their implications for the advancement of NLP techniques in understanding the complex nature of natural emotion expression. Chapter 2 illustrates the foundational theories of emotions, explaining their implications for emotion detection and defining emotion dimensions and categories. Chapter 3 describes an array of available lexicons and annotated datasets for emotion detection, particularly for English. Chapter 4 explores keyword and rule-based approaches to emotion detection, detailing methods that employ lexicons, emotion categories, and dimensions. Chapter 5 investigates supervised and unsupervised machine learning methods to detect emotion categories and dimensions in large quantity of textual data. Chapter 6 addresses the challenges and limitations of detecting emotions with neural networks and transfer learning. Finally, Chapter 7 discusses explainable AI and the ethical aspects and concerns of using Artificial Intelligence in emotion detection.

References

Adikari, A., De Silva, D., Alahakoon, D., & Yu, X. (2019). A cognitive model for emotion awareness in industrial chatbots. *IEEE 17th international conference on industrial informatics*, 183–186.

Agrawal, A., An, A., & Papagelis, M. (2020). Leveraging transitions of emotions for sarcasm detection. *Proceedings of the 43rd international ACM SIGIR conference on research and development in information retrieval* (SIGIR '20), New York, NY, USA, 1505–1508, Association for Computing Machinery.

Aman, S., & Szpakowicz, S. (2007). Identifying expressions of emotion in text. *Proceedings of the 10th international conference on text, speech, and dialogue*, 196–205.

Aman, S., & Szpakowicz, S. (2008). Using Roget's Thesaurus for fine-grained emotion recognition. *Proceedings of the third international joint conference on natural language processing*, Volume I.

Bai, Y., Ying, L., Yan, L., & Min, T. (2019). Exploiting user experience from online customer reviews for product design. *International Journal of Information Management*, 46, 173v186.

Bard, P. (1928). A diencephalic mechanism for the expression of rage with special reference to the sympathetic nervous system. *American Journal of Physiology, 84*(3), 490–516.

Barrett, L. F., Lindquist, K. A., Bliss-Moreau, E., Duncan, S., Gendron, M., Mize, J., & Brennan, L. (2007). Of Mice and Men: Natural kinds of emotions in the mammalian brain? A response to Panksepp & Izard. *Perspectives on Psychological Science, 2*, 297–312.

Barsalou, L. W. (2008). Grounded cognition. *Annual Review of Psychology, 59*, 617–645.

De Bruyne, L., De Clercq, O., & Hoste, V. (2018). LT3 at SemEval-2018 task 1: A classifier chain to detect emotions in tweets. In *Proceedings of the 12th international workshop on semantic evaluation* (pp. 123–127). Association for Computational Linguistics.

Cannon, W. B. (1927). The James-Lange theory of emotions: A critical examination and an alternative theory. *The American Journal of Psychology, 39*(1/4), 106–124.

Castillo-Sánchez, G., Marques, G., Dorronzoro, E., et al. (2020). Suicide risk assessment using machine learning and social networks: A scoping review. *Journal of Medical Systems, 44*, 205.

Christ, L., Amiriparian, S., Milling, M., Aslan, I., & Schuller, B. W. (2022). *Automatic emotion modelling in written stories. ArXiv* /abs/2212.11382

Clark, H. H. (1996). *Using language.* Cambridge University Press.

Correia Loureiro, S. M., Guerreiro, J., & Tussyadiah, I. (2021). Artificial intelligence in business: State of the art and future research agenda. *Journal of Business Research, 129*, 911–926.

Darwin, C. (1872). *The expression of emotions in man and animals.* John Murray.

Devlin, J., Chang, M-W., Lee, K., & Toutanova, K. (2019). *BERT: Pre-training of deep bidirectional transformers for language understanding. Arxiv,* https://doi.org/10.48550/arXiv.1810.04805

Ekman, P., & Friesen, W. V. (1978). *Manual for the facial action code.* Consulting Psychologist Press.

Ekman, P., & Friesen, W. V. (1971). Constants across cultures in the face and emotion. *Journal of Personality and Social Psychology, 17*, 124–129.

ElSherief, M., Ziems, C., Muchlinski, D., Anupindi, V., Seybolt, J., De Choudhury, M., & Yang, D. (2021). Latent hatred: A benchmark for understanding implicit hate speech. In *Proceedings of the 2021 conference on empirical methods in natural language processing* (pp. 345–363). Association for Computational Linguistics.

Fadhil, A., Yunlong, W., & Reiterer, H. (2019). Assistive conversational agent for health coaching: A validation study. *Methods of Information in Medicine, 58*(1), 9–23.

Fitzpatrick, K. K., Darcy, A., & Vierhile, M. (2017). Delivering cognitive behavior therapy to young adults with symptoms of depression and anxiety using a fully automated conversational agent (Woebot): A randomized controlled trial. *JMIR Mental Health, 4*(2), e19.

Frijda, N. H. (1986). *The emotions.* Cambridge University Press.

Ghazi, D., Inkpen, D., & Szpakowicz, S. (2010). Hierarchical versus flat classification of emotions in text. In *Proceedings of the NAACLHLT 2010, workshop on computational approaches to analysis and generation of emotion in text* (pp. 140–146), Association for Computational Linguistics.

Ghazi, D., Inkpen, D., & Szpakowicz, S. (2012). Prior versus contextual emotion of a word in a sentence. In *Proceedings of the 3rd workshop in computational approaches to subjectivity and sentiment analysis,* (pp. 70–78). Association for Computational Linguistics.

Giuntini, F.T., Cazzolato, M.T., dos Reis, M.d.J.D. et al. (2020). A review on recognizing depression in social networks: Challenges and opportunities. *Journal of Ambient Intelligence and Humanized Computing, 11*, 4713–4729.

Gupta, M., Bansal, A., Jain, B., Rochelle J., Oak, A., & Jalali, M.S. (2021). Whether the weather will help us weather the COVID-19. Using machine learning to measure Twitter users' perceptions. *International Journal of Medical Informatics, 145*, 104340.

Hien, H.T., Cuong, P., Nam, L.N., Nhung, H.L., & Dinh, T.L. (2018). Intelligent assistants in higher-education environments: The FIT-EBot, a chatbot for administrative and learning support. In *Proceedings of the 9th international symposium on information and communication technology.*

Hsu, C. T., Jacobs, A. M., Citron, F. M., & Conrad, M. (2015). The emotion potential of words and passages in reading Harry Potter–an fMRI study. *Brain and Language, 142*, 96–114.

Huisman, S. M., Kraiss, J. T., & de Vos, J. A. (2024). Examining a sentiment algorithm on session patient records in an eating disorder treatment setting: A preliminary study. *Frontiers in Psychiatry, 15.*

Inkster, B., Sarda, S., & Subramanian, V. (2018). An empathy-driven, conversational artificial intelligence agent (Wysa) for digital mental well-being: Real-world data evaluation mixed-methods study. *JMIR mHealth and uHealth, 6*(11), e12106.

Izard, C. E. (2007). Basic emotions, natural kinds, emotion schemas, and a new paradigm. *Perspectives on Psychological Science, 2*, 260–280.

Jafari, A. R., Li, G., Rajapaksha, P., Farahbakhsh, R., & Crespi, N. (2023). Fine-grained emotions influence on implicit hate speech detection. *IEEE Access, 11*, 105330–105343.

Lange, C.G. (1887) *Ueber Gemüthsbewgungen. Eine Psycho-Physiologische Studie*. Theodor Thomas.

Ma, C., Prendinger, H., & Ishizuka, M. (2005). Emotion estimation and reasoning based on affective textual interaction. In J. Tao, T. Tan, & R. W. Picard (Eds.) *Affective computing and intelligent interaction*. Lecture Notes in Computer Science, 3784, Springer.

Mahajan, R., & Zaveri, M. (2020). Humor identification using affect-based content in target text. *Journal of Intelligent & Fuzzy Systems, 39*(1), 697–708.

Martin, R. A., & Ford, T. E. (2018). *The Psychology of humor: An integrative approach* (2nd Edition). Elsevier Academic Press.

Martins, R., Gomes, M., Almeida, J. J., Novais, P. & Henriques, P. (2018). Hate speech classification in social media using emotional analysis. In *7th Brazilian conference on intelligent systems (BRACIS)* (pp. 61–66) . Sao Paulo, Brazil.

Mikolov, T., Chen, K., Corrado, G., & Dean, J. (2013). *Efficient estimation of word representations in vector space*. ArXiv https://doi.org/10.48550/arXiv.1301.3781

Mirtalaie, M. A., Hussain, O. K., Chang, E., & Hussain, F. K. (2018). Extracting sentiment knowledge from pros/cons product reviews discovering features along with the polarity strength of their associated opinions. *Expert Systems with Applications, 114*, 267–288.

Mohammad, S. (2018). Obtaining reliable human ratings of valence, arousal, and dominance for 20,000 English words. In *Proceedings of the 56th annual meeting of the association for computational linguistics* (Vol. 1, pp. 174–184).

Mohammad, S. (2020). Sentiment analysis: Automatically detecting valence, emotions, and other affectual states from text. *Emotion Measurement* (2nd Edition), 323–379.

Mohammad, S. (2012). From once upon a time to happily ever after: tracking emotions in mail and books. *Decision Support Systems, 53*(4), 730–741.

Mohammad, S., & Turney, P. (2013). Crowdsourcing a word-emotion association lexicon. *Computational Intelligence, 29*(3), 436–465.

Montenegro, J. L. Z., da Costa, C. A., & da Rosa Righi, R. (2019). Survey of conversational agents in health. *Expert Systems with Applications: An International Journal, 129*, 56–67.

Pacheco-Lorenzo, M. R., Valladares-Rodríguez, S. M., Anido-Rifón, L. E., & Fernández-Iglesias, M. J. (2021). Smart conversational agents for the detection of neuropsychiatric disorders: A systematic review. *Journal of Biomedical Information, 113*, 103632.

Perikos, I., & Hatzilygeroudis, I. (2013). Recognizing emotion presence in natural language sentences. In L. Iliadis, H. Papadopoulos, & C. Jayne (Eds.) *Engineering applications of neural networks* (pp. 30–39).

Pool, C., & Nissim, M. (2016). *Distant supervision for emotion detection using Facebook reactions*. ArXiv https://doi.org/10.48550/arXiv.1611.02988

Provoost, S., Ruwaard, J., van B., W., Riper, H. & Bosse, T. (2019). Validating automated sentiment analysis of online cognitive behavioral therapy patient texts: An exploratory study. *Frontiers in Psychology*, 10.

Reagan, A. J., Mitchell, L., Kiley, D., Danforth, C. M., & Dodds, P. S. (2016). The emotional arcs of stories are dominated by six basic shapes. *EPJ Data Science, 5*(1), 1–12.

Ríssola, E. A., Losada, D. E., & Crestani, F. (2021). A survey of computational methods for online mental state assessment on social media. *ACM Transaction on Computing for Healthcare, 2*(2), 1–31.

Roget, P. M. (1852). *Roget's Thesaurus of English Words and Phrases*. Longman Group Limited.

Russell, J. A. (2003). Core affect and the psychological construction of emotion. *Psychological Review, 110*(1), 145–172.

Safi Samghabadi, N., Patwa, P., Pykl, S., Mukherjee, P., Das, A., & Solorio, T. (2020). Aggression and misogyny detection using BERT: A multi-task approach. In *Proceedings of the second workshop on trolling, aggression and cyberbullying* (pp. 126–131). European Language Resources Association.

Sailunaz, K., & Alhajj, R. (2019). Emotion and sentiment analysis from Twitter text. *Journal of Computational Science, 36*, 101003.

Sailunaz, K., Dhaliwal, M., Rokne, J., & Alhajj, R. (2018). Emotion detection from text and speech: A survey. *Social Network Analysis and Mining, 8*, 28.

Sawhney, R., Joshi, H., Gandhi, S., & Shah, R. R. (2020). A time-aware transformer-based model for suicide ideation detection on social media. In *Proceedings of the 2020 conference on empirical methods in natural language processing* (pp. 7685–7697). Association for Computational Linguistics.

Scherer, K.R. (2001). Appraisal considered as a process of multilevel sequential checking. In K.R. Scherer, A. Schorr, & T. Johnstone (Eds.), *Appraisal Processes in Emotion: Theory, Methods, Research*, 94, Oxford University Press.

Scherer, K. R., & Wallbott, H. G. (1994). Evidence for universality and cultural variation of differential emotion response patterning. *Journal of Personality and Social Psychology, 66*(2), 310–328.

Shaikh, A., Peprah, E., Mohamed, R. H., Asghar, A., Andharia, N. V., Lajot, N. A., & Hussain Qureshi, M. F. (2021). COVID-19 and mental health: A multi-country study—the effects of lockdown on the mental health of young adults. *Middle East Current Psychiatry, 28*(1), 51.

Shivhare, S. N., Garg, S., & Mishra, A. (2015). Emotionfinder: detecting emotion from blogs and textual documents. In *International conference on computing, communication & automation* (ICCCA), (pp. 52–57).

Smith, C. A., & Lazarus, R. S. (1993). Appraisal components, core relational themes, and the emotions. *Cognition & Emotion, 7*(3–4), 233–269.

Solomon, R., & Stone, L. (2002). On "positive" and "negative" emotions. *Journal for the Theory of Social Behaviour, 32*, 417–435.

Song, D., Oh, E. Y., & Rice, M. (2017). Interacting with a conversational agent system for educational purposes in online courses. In *10th International conference on human system interactions* (pp. 78–82).

Suhasini, M, & Srinivasu, B. (2020). Emotion detection framework for Twitter data using supervised classifiers. Data engineering and communication technology. *Advances in Intelligent Systems and Computing*, 565–576.

Sun, Z., Han, L., Huang, W., Wang, X., Zeng, X., Wang, M., & Yan, H. (2015). Recommender systems based on social networks. *Journal of Systems and Software, 99*, 109–119.

Tadesse, M. M., Lin, H., Xu, B., et al. (2019). Detection of depression-related posts in Reddit social media forum. *IEEE Access, 7*, 44883–44893.

Tao, J. (2004). Context-based emotion detection from text input. In *Proceedings of the 8th international conference on spoken language processing* (pp. 1337–1340).

Tyng, C. M., Hafeez, A. U., Sahad, N. M., & Aamir, S. M. (2017). The influences of emotion on learning and memory. *Frontiers in Psychology, 8*.

Valitutti, A., Strapparava, C., & Stock, O. (2004). Developing Affective Lexical Resources. *Psychnology Journal, 2*, 61–83.

Vaswani, A., Shazeer, N., Parmar, N., Uszkoreit, J., Jones, L., Gomez, A. N., Kaiser, & Polosukhin, I. (2017). Attention is all you need. In *Proceedings of the 31st international conference on neural information processing systems.*

Wang, D., Li, J., Xu, K., & Wu, Y. (2017). Sentiment community detection: Exploring sentiments and relationships in social networks. *Electronic Commerce Research, 17*(1), 103–132.

Watson, L., Jurek-Loughrey, A., Devereux, B. & Murphy. B. (2020). Does history matter? Using narrative context to predict the trajectory of sentence sentiment. In *Proceedings of the second workshop on linguistic and neurocognitive resources*, Marseille, France, 8–42, European Language Resources Association.

Whissell, C. (2003). Readers' opinions of romantic poetry are consistent with emotional measures based on the dictionary of affect in language. *Perceptual and Motor Skills, 96*(3), 748–754.

Widen, S. C., & Russell, J. A. (2003). A closer look at preschoolers' freely produced labels for facial expressions. *Developmental Psychology, 39*(1), 114–128.

William, J. (1884). What is an emotion? *Mind, 9*, 188–205.

Yasmina, D., Hajar, M., & Hassan, A. (2016). Using YouTube comments for text-based emotion recognition. *Procedia Computer Science, 83*, 292–299.

Theoretical Foundations of Emotions

2

2.1 Theories of Emotion and Their Implications for Emotion Detection

Human emotions have been explored, defined, sorted, and analysed by specialists from various disciplines, encompassing psychology, sociology, neuroscience, and biology. The initial journeys into emotion exploration were embedded in philosophy and physiology. While the James-Lange (James, 1884; Lange, 1887) and the Cannon-Bard (Cannon, 1927; Bard, 1928) theories essentially describe emotions from a physiological standpoint, Darwin (1872) was the first to emphasise the evolutionary and communicative functions of emotions. Following Darwin's foundational work, cultural anthropologists Margaret Mead (1953, 1975) and psychologist Paul Ekman (Ekman et al., 1969) significantly advanced Darwin's contributions to emotion communication. Mead emphasised the cultural variability of emotional expressions, arguing that emotions are evolutionary, biologically grounded, and, at the same time, shaped by societal norms and cultural practices. In contrast, Ekman identified six basic emotions—happiness, sadness, fear, disgust, anger, and surprise—which, he argued, were universally recognisable across different cultures, supporting the idea that these emotional expressions are innate and hardwired in the way they are expressed. This observation was further nuanced by Ekman (1972) with the concept of display rules for emotions, which are culturally specific norms that dictate the appropriate expressions of emotions in a social context. Display rules may require diminishing, exaggerating, concealing entirely, or masking the emotions experienced.

Russell critiqued his view of emotions, arguing against the universality of discrete emotions and proposing instead the circumplex model of affect, suggesting that emotions are distributed in a two-dimensional circular space comprising the arousal and valence dimensions (Russell, 1980, 2003). From a neuroscience perspective, basic emotions are critiqued

F. Cavicchio, *Emotion Detection in Natural Language Processing*, Synthesis Lectures on Human Language Technologies, https://doi.org/10.1007/978-3-031-72047-5_2

by emphasising the complexity of emotional processes in the brain, suggesting that emotions cannot be neatly categorised into discrete units (LeDoux, 1996, 2012). Davidson focused on the neural bases of emotion and proposed that emotional processes are more varied and context-dependent than Ekman's model suggests (Davidson, 1992, 2003). Similarly, Damasio (1999) integrated the earlier biological theories of emotions with insights into the neural underpinnings of emotions, acknowledging both innate mechanisms and the influence of context and experience.

Similarly, Lisa Feldman Barrett challenged the universality and innateness of basic emotions, arguing that emotions are not hardwired, discrete entities but are constructed through the integration of sensory input, prior experiences, and contextual information. Her theory posits that emotions are flexible, context-dependent constructs rather than fixed, biologically determined responses (Barrett, 2006, 2017). Barrett suggests that what we perceive as distinct emotions are, in fact, complex psychological states created by the brain's predictive mechanisms. Her view suggests that emotions are constructed from more fundamental psychological ingredients, such as core affect and conceptual knowledge, tailored to the specific demands of the situation (Barrett, 2011, 2013).

The debate on emotion theory illustrates the complexity of understanding emotions. While early theories posited biological and universal bases for emotions, later research has highlighted the significant role of context and cultural factors in shaping emotional experiences, expression and communication.

As we have seen in Chap. 1, concepts associated with emotions such as affect, feeling, emotion, sentiment, and opinion are related to human subjectivity. Subjective experiences, originating from an individual's perspective and reflecting their desires, beliefs, and feelings, make emotion detection a complex NLP task and a fascinating research pursuit. Current initiatives in text-based emotion detection demonstrate constraints in detection proficiencies. For instance, most sentiment analysis and opinion mining research mainly focus on identifying text polarity (positive, negative, and neutral), albeit sentiments, opinions, and emotions harbour complexities that surpass simple polarity. Affects, feelings, emotions, and sentiments are often understood similarly and used interchangeably in NLP research. In the following, we try to illuminate the various facets of these terms.

Traditionally, *affect* denoted the sensation of experiencing a feeling. In contemporary psychology terminology, affect pertains to the mental equivalent of internal bodily representations linked with emotions and actions. Shouse (2005) described affect as a non-conscious experience of intensity and the body's way of preparing for action by adding quantitative intensity to the quality of an experience. Thus, affect may be defined as the positive or negative evaluation of an entity, idea, action, or object, incorporating the intensity dimension. Affect plays a central role across various psychological phenomena including emotions (Barrett, 2006; Diener et al., 1999), attitudes (e.g., Cacioppo & Berntson, 1994; Eagly & Chaiken, 1998; Ito & Cacioppo, 2005), but also stereotyping and prejudice (Cacioppo & Berntson, 2001), negotiation strategies (Forgas, 1998), moral judgements and decision-making (Haidt, 2002), future predictions (Gilbert & Ebert,

2002), and personality (Yik et al., 2002). Affect is foundational in language comprehension, influencing speakers' vocal characteristics and acoustical features (Nygaard & Lunders, 2002) and affecting lexical processing (Schirmer & Kotz, 2003), simplifying or complicating word recognition (Nygaard & Queen, 2008).

According to the APA Dictionary of Psychology, a *feeling is* a standalone occurrence. Feelings embody subjective and evaluative sensations that are assessed as pleasant or unpleasant. Feelings can also exhibit more specific characteristics; for instance, the tone distinguishable in fear is perceived differently from that in anger. The main feature that separates feelings from cognitive, sensory, or perceptual experiences is their association with appraisal. In contrast to emotions, feelings are purely mental, whereas emotions are formulated and expressed in interaction with the external world. Shouse (2005) defined a *feeling* as a sensation related to and labelled according to past experiences. Feelings are intrinsically biographical since each person has a distinct reservoir of past sensations to draw upon when interpreting and labelling their feelings. For Wierzbicka (1995) feelings are universal, as every language appears to have a term for 'feel'.

In the psychology literature, we find very different definitions of *emotion*. For example, following Izard (2010), emotions can be categorised into two general types: fundamental emotions, which are episodical, and emotions characterised by a dynamic interaction with cognition. Scherer (2000) defined *emotion* as "episodes of synchronised alterations in several components (including neurophysiological activation, motor expression, and subjective feeling, but also possibly action tendencies and cognitive processes) in response to internal or external events of significant importance to the organism". Scherer warned against equating affective valence with emotional episodes or affective states, such as attitudes. Nonetheless, emotions, as a form of affective state, indeed possess valence and intensity. Thoits (1989) identified emotions as culturally determined feelings or effects, and Shouse (2005) conveyed that emotional expression can simultaneously reflect our internal state and adhere to social expectations and norms. Batson et al. (1997) further characterised emotions as reflections of the presence of a specific goal or perceived shifts in relationship to a specific goal.

Cattell (2006) described *sentiment* as a motivational structure that is developed by learning about entities, actions, and ideologies. A sentiment can be reinforced or changed over time, and it can elicit strong positive or negative responses. Broad (1954) stated that sentiments are formed by consistently associating an emotional tendency with an entity and can manifest as various emotions, such as love, hate, or anger. However, in NLP, the term sentiment seems to be employed in the same sense of economic and consumer behaviour. Pioneers of consumer behaviour and behavioural economy, such as Katona (1975), Simon (1955), and Tversky and Kahneman (1973), championed the significance of motives, attitudes, and expectations in influencing economic activities, challenging traditional economic theories of rational economic human behaviour. With surveys on consumer attitudes, a new era of evaluating consumer perceptions began, scrutinising their sentiment towards goods in terms of positive, negative, or neutral assessments.

To summarise, the main distinctive factors among affect, feelings, emotions, and sentiment are the following:

- **Affect** precedes both feelings and emotions and is largely non-conscious.
- **Feelings** are conscious phenomena centred on the individual and their phenomenological experiences, they express affect.
- **Emotions** express affect and feelings, and they are influenced by culture. They can be episodical (fundamental emotions) and described in terms of their valence and intensity.
- **Sentiment** is a continuous emotional disposition towards an external entity developed over time.

From the standpoint of psychology and its studies on emotions, existing models of emotion are classified into two primary theoretical groups: categorical and dimensional:

- **Categorical Models**: These models propose that emotions are distinct and discrete entities. According to this perspective, a limited number of basic emotions, such as happiness, sadness, anger, fear, surprise, and disgust, are universally recognized and biologically hardwired. This approach often draws on evolutionary psychology and neuroscientific research, which suggest that these primary emotions are innate and have distinct physiological and neurological correlates.
- **Dimensional Models**: Contrary to the categorical approach, dimensional models view emotions as existing along a continuum or within a multidimensional space. Standard dimensions used in these models include valence (ranging from pleasant to unpleasant) and activation/arousal (ranging from low to high). This approach emphasizes the fluidity and complexity of emotional experiences, suggesting that emotions can be understood in terms of their position within these dimensions rather than as distinct categories. This model is beneficial in explaining the subtleties and gradations of emotional experiences that do not fit neatly into predefined categories.

In Sects. 2.2 and 2.3, we delve deeper into the theories characterizing these two emotion models.

2.2 Theories of Emotion Categories

Categorical emotion models enumerate categories of emotions (e.g., happiness, sadness, etc.) each distinct from the others. Psychological scientists posit that individuals possess internal mechanisms for a restrained set of responses that can be straightforwardly and objectively measured once activated. Ekman and Friesen (1969) proposed the categorical model after researching the universal recognition of emotions from facial expressions.

They categorize emotions into six fundamental types: happiness, fear, sadness, surprise, disgust, and anger. They also appointed a very complex method to recognize emotions from micro-movements of facial muscles, the FACS (Ekman & Friesen, 1978). Additionally, Ekman and Friesen suggested that combinations of basic emotions were possible, the so-called emotion blends (Ekman & Friesen, 1975). Studies based on self-reporting suggest that individuals can often feel multiple emotions simultaneously (e.g., Oatley & Duncan, 1994; Scherer & Meuleman, 2013). The phenomenon of experiencing concurrent positive and negative states is commonly referred to as mixed emotions. A mixed emotion can be considered a specific type of blended (or compound) emotion, which involves experiencing any combination of emotions, irrespective of their positive or negative valence.

Parrot's emotion model (2000) acknowledges six fundamental emotions: fear, sadness, surprise, anger, love, and joy. Parrot expands these core categories into a comprehensive tree structure, ultimately encapsulating 100 distinct emotions. Parrot structured emotions hierarchically into three levels—primary, secondary, and tertiary—with joy, anger, fear, love, and surprise categorized as the primary set of emotions (see Fig. 2.1).

Ortony et al., (1988) stated that emotions originate from an individual's perception of events and their consequent emotions, varying in intensity degree. This model has 22

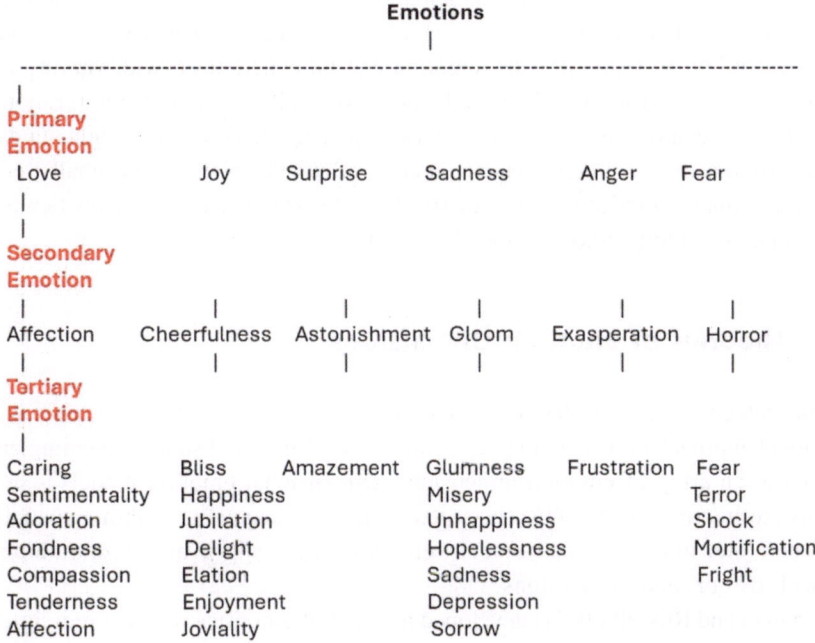

Fig. 2.1 An illustration of the hierarchy of emotions described in Parrot (2000)

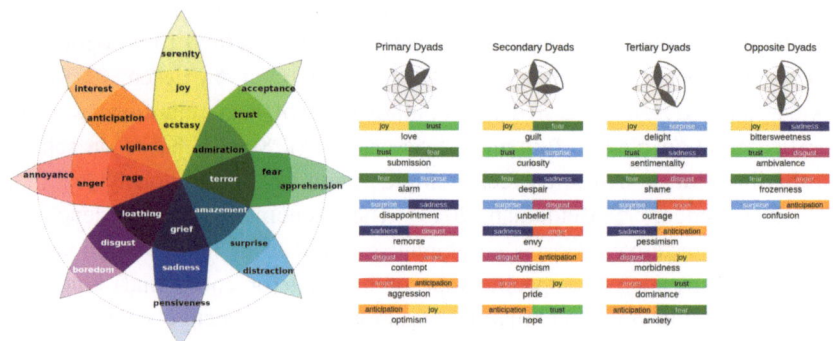

Fig. 2.2 Plutchik's emotion wheel and the primary, secondary, tertiary, and opposite dyads of emotions based on Plutchik's wheel (*Credits* Semeraro et al., 2021, pp. 6–7)

emotive categories, adding 16 emotions to Ekman and Friesen's basic ones. Ortony and colleagues' model includes emotion categories such as envy, relief, appreciation, reproach and self-reproach, shame, pity, admiration, disappointment, grief, gratification, gloating, hope, and even like and dislike.

Plutchik (1980) proposed a model involving eight primary emotions, adding to the "original six" acceptance/trust and anticipation. In Plutchik's model, emotions are in opposing pairs, such as surprise vs anticipation, joy vs sadness, anger vs fear, and trust vs disgust. According to Plutchik, each emotion can be experienced with varying intensity levels of valence and arousal, influenced by an individual's event perception. Emotions are displayed in concentric circles: the innermost featuring derivatives of eight fundamental emotions, followed by the eight fundamental emotions themselves, and finally, combinations of the primary emotions on the outermost circle. This wheel effectively demonstrates the interrelations among emotions (see Fig. 2.2).

2.3 Theories of Emotion Dimensions

While the categorical approach confines emotional states to a handful of distinct types, the dimensional approach considers the variability of the emotional states, assigning emotions according to an array of emotion dimensions. Dimensional emotion models assume that emotions are interrelated, thereby necessitating their arrangement within a spatial framework. Therefore, in dimensional models, emotions are mapped onto a multidimensional space with two or more dimensions.

Mehrabian and Russell (1974) developed a model of emotions with a three-dimensional framework encompassing pleasantness (or positivity), arousal (or responsiveness), and potency (or dominance). Emotions are differentiated by arousal (activation and deactivation) and valence (pleasantness and unpleasantness), while the third dimension,

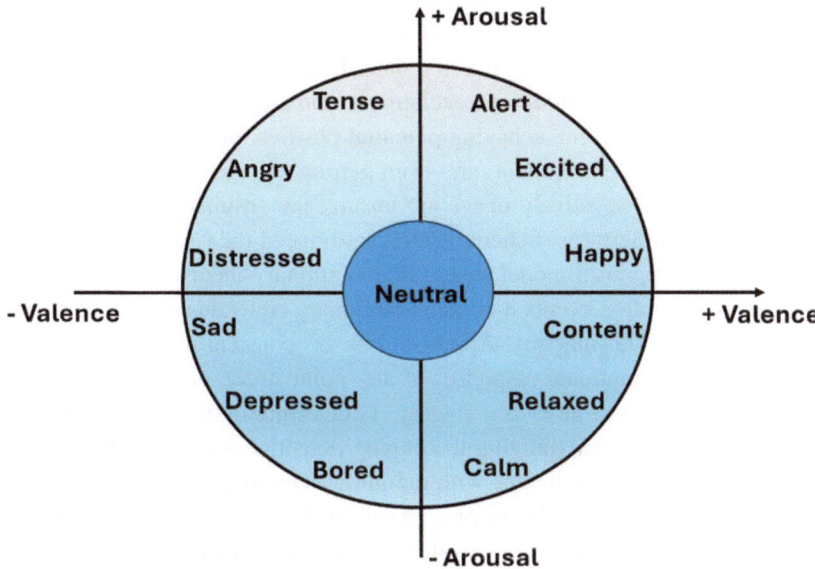

Fig. 2.3 The circumplex model of emotion with valence and arousal axes

dominance, pertains to the extent to which an individual has control over the situation or their emotions. Later, Russell (1980) introduced the circular two-dimensional model called the circumplex of affect. This model categorizes emotions within the domains of arousal and valence, with arousal distinguishing emotions through activation and deactivation and valence differentiating them through pleasantness and unpleasantness (see Fig. 2.3).

2.3.1 The Component Process Model

Appraisal theories of emotions explore why individuals exhibit varied emotive reactions to identical or similar situations. People's distinct responses to the same situation, rooted in their unique perceptions, spark various individual-specific emotions. Psychologists such as Arnold (1950) Lazarus (1991) and Scherer (2001) embarked on categorizing these emotions, laying the foundation for cognitive appraisal theory. For example, Scherer's appraisal model mandates a sequence of appraisal checks interspersed with several evaluative checks. These checks permit the inspection of stimuli in the sequential appraisal process, establishing a stepwise mechanism. According to appraisal theories, different emotions can emerge from the same event for the same subject at different time points.

Moreover, emotions can shift through components such as memory, physiology, motivation, motor response, and feelings. To explain the series of appraisal checks, Scherer et al., (2001) used the example of going on a first date. If the date is perceived positively, emotions such as happiness, joy, giddiness, excitement, and anticipation might arise, stemming from the appraisal of the event as having potential positive future implications like initiating a new relationship, commitment, and even getting engaged and married. Conversely, if the date is perceived negatively or we are unsure, the ensuing emotions might involve dejection, sadness, or alertness. Scherer (2001) introduced the Component Process Model (CPM), a cognitive appraisal model rooted in evolutionary theory. This model posits that when we appraise emotion events it is the evolutionary equivalent of when we are assessing our surroundings for survival. Appraisal can be constantly revised, and it has the potential to alter our emotional responses at any point in the process. The CPM divides the appraisal process into four check stages: (1) determining the stimulus relevance, (2) assessing its implications, (3) gauging the coping potential, and (4) evaluating its normative significance. The process begins with the initial emotional response in the first check and concludes with the fourth. As we progress through these stages, appraisal dimensions evolve from evolutionary-driven, universal evaluations to more culturally influenced ones.

Scherer calls these specific appraisal dimensions within each stage stimulus evaluation checks (or SECs). SECs are essential evaluations that we undertake to differentiate and label their emotions. Importantly, these SECs are dimensional rather than categorical, suggesting they fall on a spectrum from low to high, with an array of intermediate values. The specific emotion we identify, and its intensity are determined by the combination of the experienced SECs. Hence, to truly understand our emotional state, we should consider the collective outcome of these evaluations.

The relevance check is the initial phase of the emotional experience. As illustrated in Fig. 2.4, it begins by assessing the novelty of an event: sudden, unfamiliar changes capture attention due to their potential impact on our well-being. For instance, hearing an unexpected loud explosion might evoke fear, whereas familiar events, such as attending a lunch with colleagues, can induce a sense of contentment. After gauging the event's novelty, we evaluate the event's pleasantness, determining a generalized positive or negative feeling. According to Scherer, events deemed unpleasant generally lead to avoidance, while pleasant ones lead to approach, with anger being an exception. The relevance check concludes with a goal relevance assessment, where we judge how an event pertains to our immediate objectives, such as survival, satisfying basic needs, accomplishing a task, maintaining relationships, etc. Goal relevance operates on a spectrum. That is, events can have varying degrees of significance for us.

In the implications check, we assess how an event affects our core needs. This evaluation comprises several steps, including causal attribution, outcome probability, goal conduciveness, and urgency. Causal attribution is the process of identifying whether an event stems from our own doings (internal attribution) or external circumstances, like another individual's actions or unforeseen situations (external attribution). For example,

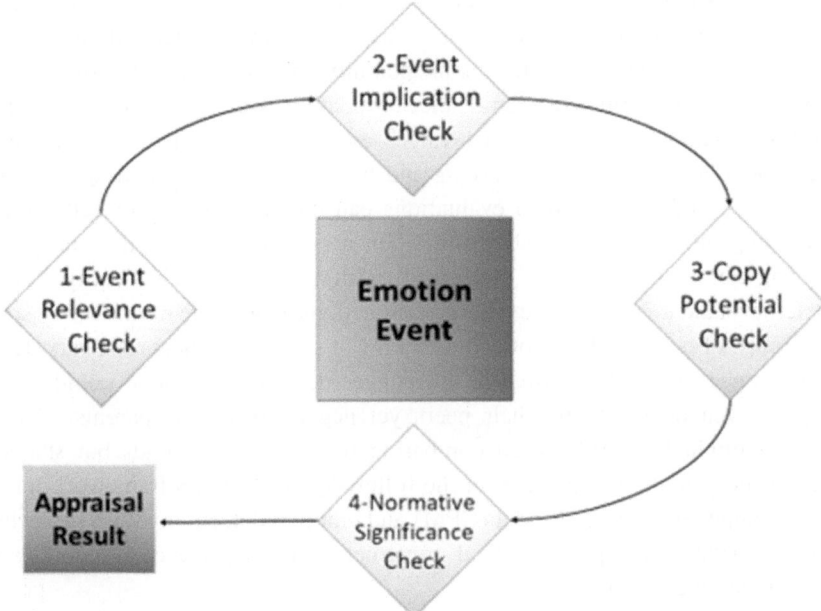

Fig. 2.4 A representation of the components process model (CPM) appraisal checks

after successfully completing a challenging project, we may feel pride due to an internal attribution. How we assign blame or credit can also shape our emotional response. After a poor performance in a contest, if we believe it was due to our lack of preparation, we might experience regret. However, if we think the contest was unfair, we are angry.

In the outcome probability check, we predict the potential outcomes of specific situations. Scherer suggests that anticipating potential reactions can shape our feelings. Imagining our teacher's dissatisfaction after a mediocre presentation might cause anxiety, but if we believe they would be impressed, we are elated. The goal conduciveness check measures whether an event's aftermath aligns with or against our goals. Meeting expectations usually leads to positive emotions such as satisfaction.

On the other hand, when results do not meet our expectations, feelings of disappointment or resentment can surface because of goal obstructiveness. The urgency check estimates the timeframe available to meet our aspirations. Feeling pressed for time corresponds to high urgency, whereas having a generous window relates to low urgency. Scherer underlines that urgency becomes particularly significant when vital objectives are at stake within a constrained timeframe.

The coping potential check is the third assessment phase. In this check, we estimate our ability to handle the emotion-triggering event. We consider possible reactions and the implications of those actions. Key components include controllability, which is the belief in our, or someone else, ability to influence an event, and power or resources, assessing if one has the means to manage the event. If altering the event's outcome is not possible, the adjustment component evaluates the adaptability and shape our emotion responses.

In the process of normative significance evaluation, we assess the events and their responses in relation to their personal values and societal standards. First, there's the internal standards check, in which events and behaviours are weighed against our self-identity, values, and morals. Such evaluations can directly impact our self-esteem. For example, lying might diminish self-worth when it contradicts our self-identity, whereas volunteering can boost confidence by aligning with our positive self-identity. Then, there is the external standards check. We measure our actions against the norms set by society or specific groups of people. The emotional outcome of such assessments can vary depending on the reference group. For instance, a teenager deciding to smoke could be viewed neutrally or even positively by their peers yet negatively by the parents. This means the teenager might feel pride when comparing themselves to friends but shame when considering the parents' perspective. In the following we sketch a framework integrating Scherer's Component Process Model (CPM) and emotion detection with NLP techniques. Our goal is bridging the psychological constructs with possible corresponding natural language processing methods:

Event Appraisal

Process: Relevance deals with how the text determines its significance to the reader
- NLP methods: *keywords, sentiment polarity* can help tracking relevance

Process: Implication is about the causality and potential outcomes in the text
- NLP method: *sentiment shifts* can help tracking the implication

Process: Coping refers to textual mechanisms suggesting how one might deal with events or situations
- NLP method: *intensifiers* can be a telltale sign of coping

Process: Normative significance examines how the text aligns (or misaligns) with personal or societal standards.
- NLP method: *tracking cultural references, moral references, norms, and taboos* can help tracking normative significance

Cognitive Mechanism

Process: Attention and memory encompass which parts of the text are noteworthy and memorable.
- NLP method: *salient terms, recurring themes* usually mark salient, memorable events

Process: Motivation seeks to understand what drives responses through the text.
- NLP method: *call-to-actions, questions* can pinpoint the motivation

Process: Reasoning explores logical analysis.

- NLP method: *argument structures, claims, evidence* are evident signs of reasoning
Process: Self involves the promotion of introspection or self-reflection in the text.
- NLP method: *personal anecdotes, reflective questions* can help detecting the self

Behavioural and Physiological Response

Process: Appraisal processes evaluate events in the text.
- NLP method: *sentiment analysis, emotion lexicons* can be results of the appraisal process
Process: Autonomic Physiology. Since we cannot measure the autonomic physiology response of the writers, cues of calmness as well as intensity markers in the text are crucial.
- NLP method: *arousal, intensity markers*
Process: Action tendencies hint at an inclination towards specific behaviours.
- NLP method: *imperatives, suggestions* can manifest action tendencies in the text.
Process: Motor Expression. We cannot see the motor expressions of the writers. However, we can speculate on the potential physical reactions through textual markers.
- NLP method: *markers of excitement, surprise, shock* can express motor expressions
Process: Subjective feeling captures the personal emotional experiences of the writer.
- NLP method: *first-person accounts, emotion adjectives and adverbs* can express subjective feelings

2.4   Conclusion

In this chapter, we have explored the multifaceted landscape of emotion theories. We have explored various models of emotion, compared the categorical and dimensional approaches, and introduced the circumplex model of affect and Scherer's Component Process Model (CPM). These models emphasise the multidimensional nature of emotions and the cognitive appraisal processes that underlie them, illustrating the broad spectrum and complex combinations of human emotions.

We have also proposed a framework aimed at integrating the theoretical concepts of CPM with practical NLP techniques for detecting emotions in texts. The bridging of theory and NLP methods not only enhances our understanding of emotion theories but also suggests the way for advanced applications in the field of NLP.

The importance of incorporating these emotion theories into NLP methods cannot be overstated. By leveraging these models, we can better disambiguate the contextual information of the emotions expressed in texts, leading to more accurate and sophisticated NLP systems. This integration is crucial for improving the reliability and validity of emotion detection algorithms. Considering these theories ensures that our technological approaches are grounded in a comprehensive understanding of human emotions, fostering advancements that are both methodologically and practically impactful.

References

Arnold, M. B. (1950). *Emotion and personality* (Vol. 1). Columbia University Press.

Bard, P. (1928). A diencephalic mechanism for the expression of rage with special reference to the central nervous system. The American Journal of Physiology, 84(3), 490-515. https://doi.org/10. 1152/ajplegacy.1928.84.3.490

Barrett, L. F. (2006). Are emotions natural kinds? *Perspectives on Psychological Science, 1*(1), 28–58.

Barrett, L. F. (2011). Constructing emotion. *Psychological Topics, 20*(3), 359–380.

Barrett, L. F. (2013). Psychological construction: The DARPA challenge. *Emotion Review, 5*(4), 379–389.

Barrett, L. F. (2017). *How emotions are made: The secret life of the brain*. Houghton Mifflin Harcourt.

Batson, C. D., Early, S., & Salvarani, G. (1997). Perspective taking: Imagining how another feels versus imaging how you would feel. *Personality and Social Psychology Bulletin, 23*(7), 751–758.

Broad, C. D. (1954). Emotion and sentiment. *The Journal of Aesthetics and Art Criticism, 13*(2), 203–214.

Cacioppo, J. T., & Berntson, G. G. (1994). Relationship between attitudes and evaluative space: A critical review, with emphasis on the separability of positive and negative substrates. *Psychological Bulletin, 115*(3), 401–423.

Cacioppo, J. T., & Berntson, G. C. (2001). The affect system and racial prejudice. In J. A. Bargh & D. K. Apsley (Eds.), *Unravelling the complexities of social life: A festschrift in honour of Robert B. Zajonc* (pp. 95–110). American Psychological Association.

Cattell, R. (2006). Sentiment or attitude? The core of a terminology problem in personality research. *Journal of Personality, 9*, 6–17.

Cannon, W. B. (1927). The James-Lange theory of emotions: A critical examination and an alternative theory. *The American Journal of Psychology, 39*(1/4), 106–124. https://doi.org/10.2307/141 5404

Damasio, A. (1999). *The feeling of what happens: Body and emotion in the making of consciousness*. Harcourt College Publishers.

Darwin, C. (1872). *The expression of the emotions in man and animals*. John Murray.

Davidson, R. J. (1992). Emotion and affective style: Hemispheric substrates. *Psychological Science, 3*(1), 39–43.

Davidson, R. J. (2003). Affective neuroscience and psychophysiology: Toward a synthesis. *Psychophysiology, 40*(5), 655–665.

Diener, E., Suh, E. M., Lucas, R. E., & Smith, H. L. (1999). Subjective well-being: Three decades of progress. *Psychological Bulletin, 125*(2), 276–302.

Eagly, A. H., & Chaiken, S. (1998). Attitude structure and function. In D. T. Gilbert, S. T. Fiske, & G. Lindzey (Eds.), *The handbook of social psychology* (pp. 269–322). McGraw-Hill.

Ekman, P., & Friesen, W. V. (1975). *Unmasking the face: A guide to recognizing emotions from facial clues*. Prentice-Hall.

Ekman, P., & Friesen, W. V. (1978). *Manual for the facial action code*. Consulting Psychologist Press.

Ekman, P. (1972). Universals and cultural differences in facial expressions of emotion. In J. Cole (Ed.), *Nebraska symposium on motivation, 19* (pp. 207–282). University of Nebraska Press.

Ekman, P., & Friesen, W. V. (1969). The repertoire or nonverbal behavior: Categories, origins, usage and coding. *Semiotica, 1*, 49–98.

Ekman, P., Sorenson, E. R., & Friesen, W. V. (1969). Pan-cultural elements in facial displays of emotions. *Science, 164*(3875), 86–88.

Forgas, J. P. (1998). On feeling good and getting your way: Mood effects on negotiator cognition and bargaining strategies. *Journal of Personality and Social Psychology, 74*(3), 565–577.

Gilbert, D. T., & Ebert, J. E. J. (2002). Decisions and revisions: The affective forecasting of changeable outcomes. *Journal of Personality and Social Psychology, 82*(4), 503–514.

Haidt, J. (2002). Dialogue between my head and my heart: Affective influences on moral judgment. *Psychological Inquiry, 13*(1), 54–56.

Ito, T., & Cacioppo, J. (2005). Variations on a human universal: Individual differences in positivity offset and negativity bias. *Cognition & Emotion, 19*, 1–26.

Izard, C. E. (2010). The many meanings/aspects of emotion: Definitions, functions, activation, and regulation. *Emotion Review, 2*(4), 363–370.

James, W. (1884). What is an emotion? *Mind, 9*(34), 188–205. https://doi.org/10.1093/mind/os-IX.34.188

Katona, G. (1975). *Psychological economics.* Elsevier.

Lange, C. (1887). The mechanism of the emotions. In B. Rand (Ed.), *The classical psychologists* (pp. 672–684). Houghton Mifflin.

Lazarus, R. S. (1991). *Emotion and adaptation.* Oxford University Press.

LeDoux, J. E. (1996). *The emotional brain: The mysterious underpinnings of emotional life.* Simon & Schuster.

LeDoux, J. E. (2012). Rethinking the emotional brain. *Neuron, 73*(4), 653–676.

Mead, M. (1953). The study of culture at a distance. In M. Mead & R. Metraux (Eds.), *The study of culture at a distance* (pp. 3–53). University of Chicago Press.

Mead, M. (1975). Review of Darwin and facial expression. *Journal of Communication, 25*, 209–213.

Mehrabian, A., & Russell, J. A. (1974). *An approach to environmental psychology.* MIT Press.

Nygaard, L. C., & Lunders, E. R. (2002). Resolution of lexical ambiguity by emotional tone of voice. *Memory & Cognition, 30*(4), 583–593.

Nygaard, L. C., & Queen, J. S. (2008). Communicating emotion: Linking affective prosody and word meaning. *Journal of Experimental Psychology: Human Perception and Performance, 34*(4), 1017–1030.

Oatley, K., & Duncan, E. (1994). The experience of emotions in everyday life. *Cognition & Emotion, 8*(4), 369–381.

Ortony, A., Clore, G. L., & Collins, A. (1988). *The cognitive structure of emotions.* Cambridge University Press.

Parrot, W. (2000). *Emotions in social psychology.* Psychology Press.

Plutchik, R. (1980). A general psychoevolutionary theory of emotion. In R. Plutchik & H. Kellerman (Eds.), *Theories of emotion* (pp. 3–33). Academic press.

Russell, J. A. (1980). A circumplex model of affect. *Journal of Personality and Social Psychology, 39*(6), 1161–1178.

Russell, J. A. (2003). Core affect and the psychological construction of emotion. *Psychological Review, 110*(1), 145–172.

Scherer, K. R., Shorr, A., & Johnstone, T. (Ed.). (2001). *Appraisal processes in emotion: theory, methods, research.* Oxford University Press.

Scherer, K.R., & Meuleman, B. (2013). Human emotion experiences can be predicted on theoretical grounds: evidence from verbal labelling. *PLoS One, 8*(3).

Scherer, K. R. (2000). Psychological models of emotion. In J. C. Borod (Ed.), *The neuropsychology of emotion* (Vol. 137, pp. 137–162). Oxford University Press.

Scherer, K. R. (2001). Appraisal considered as a process of multilevel sequential checking. In K. R. Scherer, A. Schorr, & T. Johnstone (Eds.), *Appraisal processes in emotion: Theory, methods, research* (pp. 92–120). Oxford University Press.

Schirmer, A., & Kotz, S. A. (2003). ERP evidence for a sex-specific Stroop effect in emotional speech. *Journal of Cognitive Neuroscience, 15*(8), 1135–1148.

Semeraro, A., Vilella, S., & Ruffo, G. (2021). PyPlutchik: Visualising and comparing emotion-annotated corpora. *PLoS ONE, 16*(9), e0256503.

Shouse, E. (2005). Feeling, emotion, affect. *M/C Journal, 8*(6).

Simon, H. A. (1955). A behavioural model of rational choice. *The Quarterly Journal of Economics, 69*(1), 99–118.

Thoits, P. A. (1989). The sociology of emotions. *Annual Review of Sociology, 15*, 317–342.

Tversky, A., & Kahneman, D. (1973). Availability: A heuristic for judging frequency and probability. *Cognitive Psychology, 5*(2), 207–232.

Wierzbicka, A. (1995). The relevance of language to the study of emotions. *Psychological Inquiry, 6*(3), 248–252.

Yik, M., Russell, J. A., Ahn, C.-K., Fernandez Dols, J. M., & Suzuki, N. (2002). Relating the Five-Factor Model of personality to a circumplex model of affect: A five-language study. In R. R. McCrae & J. Allik (Eds.), *The Five-Factor model of personality across cultures* (pp. 79–104). Kluwer Academic/Plenum Press Publishers.

Resources for Emotion Detection: Lexicons and Annotated Datasets

3

3.1 Emotion Lexicons and Annotated Datasets

This chapter explores the tools used for emotion detection in Natural Language Processing (NLP). A comprehensive understanding of the critical instruments and resources underpinning this research area is essential for grasping the field's development. In the following, we provide an overview of the lexicons and annotated datasets that have been foundational in detecting emotions in textual data, focusing on the early advancements that shaped the field of emotion detection in NLP. These resources will be referenced throughout Chaps. 4, 5, And 6 since they are still relevant for more recent Artificial Intelligence (AI) models for emotion detection.

Emotion detection in NLP relies heavily on high-quality data. Annotated datasets, meticulously labelled with emotional tags, are indispensable for rule-based systems, in which such resources facilitate the preprocessing, feature extraction, and analysis necessary to achieve accurate emotion classification. Emotion lexicons and annotated datasets are also essential for training and evaluating AI models. Indeed, these datasets provide the foundational data needed to develop and test various emotion detection algorithms. By laying out the fundamental resources in this chapter, we aim to provide the necessary context to appreciate the advancements and methodologies discussed in the following sections.

Lexicons and annotated datasets each offer unique advantages for rule-based emotion detection. Lexicons provide a straightforward mapping of words to emotions, allowing for quick and direct classification. Annotated datasets, with their contextual depth, facilitate the development of rules that capture the subtleties and complexities of how emotions are expressed in various contexts. Regarding machine learning and deep learning techniques and models for emotion detection, lexicons provide a straightforward mapping of words to emotions, offering a valuable resource for feature extraction and initial model training.

© The Author(s), under exclusive license to Springer Nature Switzerland AG 2025   33
F. Cavicchio, *Emotion Detection in Natural Language Processing*, Synthesis Lectures on Human Language Technologies, https://doi.org/10.1007/978-3-031-72047-5_3

Annotated datasets enable machine learning and deep learning models to learn contextual information from the examples in the dataset. The use of emotion lexicons and annotated datasets to capture the subtleties and complexities of emotional expression in various contexts leads to more accurate and sophisticated emotion detection.

In the following sections, we will discuss the foundational lexicons and annotated datasets used in categorical and dimensional emotion detection. We will present the lexicons and annotated datasets that primarily target the detection of emotion categories. Subsequently, we will delve into the lexicons and annotated datasets designed to detect emotion dimensions. While lexical resources exist for many languages, and some are multilingual, our primary focus in this context will be on resources specifically designed for the English language.

3.1.1 Lexicons

An emotion lexicon is a repository of words, phrases, or patterns capturing the diverse spectrum of human emotions. By differentiating between direct and indirect affective words, emotion lexicons can offer a context-sensitive understanding of the text, enabling a more accurate sentiment analysis and emotion detection. For example, with the help of lexicons, we can classify words or phrases into broader emotional categories. Lexicons can be built to classify the text with specific emotions, such as *Joy*, *Sadness*, *Anger*, and *Surprise*. Lexicons can also have different granularity, pinpointing specific emotional states. These lexicons can provide deeper insights into a text's exact emotional nuance: for instance, instead of identifying *Sadness*, they might differentiate between *Melancholy*, *Grief*, and *Despair*.

3.1.2 Annotated Datasets

In emotion research, popular models such as Paul Ekman's six basic emotions (Ekman & Friesen, 1969), which categorise emotions into distinct classes such as *Happiness* and *Anger*, often form the bedrock of annotation. On the other hand, dimensional models, such as Mehrabian and Russell's (1974) circumplex model of affect, depict emotions across continuous dimensions (e.g., valence and arousal). Some annotated datasets annotate emotions beyond the basic ones and try to capture emotions as specific as *Enthusiastic* or *Melancholic*. The annotation granularity usually mirrors the depth and detail of the emotion detection sought by the study. How emotions are annotated can differ, too. For example, in annotated datasets with dimensional annotation, the intensity of an emotion can be gauged on a scale, providing a further layer of detail in the annotation. Multiple annotators often collaborate in the annotation of the annotated datasets. Sometimes, annotators are experts, for example, psychologists. In other cases, a vast and varied pool of crowd-sourcing annotators are recruited through online websites such as Amazon's Mechanical Turk or CrowdFlower.

A crucial point of annotated datasets annotations is the annotation reliability across annotators. Measures such as the kappa statistics (Artstein & Poesio, 2008) are commonly used to ensure that annotators agree on the emotion labels attributed to the text above the chance level. Kappa statistics are commonly used in statistics and research to measure inter-annotator agreement for qualitative items. They quantify whether agreement among coding raters exceeds what would be expected by chance. The basic form of kappa statistics is Cohen's kappa, which is designed for two raters. The Cohen's kappa formula is the following:

$$\kappa = \frac{P_0 - P_e}{1 - P_e}$$

where P_0 is the relative observed agreement among raters (the proportion of times the raters agree) and P_e is the hypothetical probability of chance agreement.

When there are more than two annotators, Fleiss's kappa is used. Fleiss's kappa extends the idea to any fixed number of raters, assessing the reliability of agreement between multiple annotators. Fleiss's kappa is valuable because it considers the agreement occurring by chance, offering a reliable measure of the annotation consistency across multiple raters, ensuring that the results are not skewed by random agreement rate. Fleiss's kappa formula is the following:

$$\kappa = \frac{\overline{P}_0 - \overline{P}_e}{1 - \overline{P}e}$$

where $\overline{P}_0$ is the mean of the proportions of agreement for each item (i.e., the average agreement per item) and $\overline{P}_e$ is the mean of the squared proportions of each category (calculated under the assumption that the raters randomly guess categories based on the overall distribution of the categories).

Based on these statistical measures, we can refine the annotation process. Pilot studies on smaller data subsets often precede the full-fledged annotation, helping to refine guidelines and address the coding scheme ambiguities.

3.2 Lexicons for the Detection of Emotion Categories

In the following, we will describe the lexicons that have been foundational in categorical emotion detection, focusing on their design and application. Lexicons give a straightforward mapping of words to emotions, enabling quick and direct classification.

WORDNET-AFFECT (Valitutti et al., 2004) was developed starting from the WORD-NET knowledge base. Ortony et al. (1987) have highlighted the importance of distinguishing words that explicitly express emotions (e.g., *happy, angry*, etc.) from those that convey emotions contextually (e.g., *admiration, reproach,* etc.). Building on this observation, Valitutti et al. (2004) and Strapparava et al. (2006) proposed the development of an affective lexicon that aligned affective concepts with the corresponding words. They identified the WORDNET synset model as the ideal platform for developing an emotion lexicon. WORDNET is a Lexical Knowledge Base that offers structured semantic information (Miller et al., 1990). Unlike traditional dictionaries, WORDNET arranges lexical data based on meanings rather than word forms. Nouns, verbs, adjectives, and adverbs are grouped into synonym sets (or synsets), each depicting a lexical concept. We can visualise WORDNET as a lexical matrix consisting of two dimensions: one focused on lexical relations, which are language-specific, and another on conceptual relations, which should mostly transcend specific languages. The aim of WORDNET-AFFECT was to pinpoint many affective concepts, hierarchically arrange them, and interlink them through the lexicon. The starting point was a collection of psychological adjectives, notably those with affective connotations. The annotators intuitively associated nouns with adjectives, as well as verbs with adverbs. However, Valitutti and colleagues believed that relying solely on the lexicon was insufficient and that incorporating affective data from emotion models was necessary to design the affective lexicon framework. Therefore, once the core synsets were identified, they added a manually curated lexical database called AFFECT. AFFECT contains 1,903 terms that touch upon mental states, such as *cheerful, cheer up,* and *cheerfully* (Strapparava et al., 2006). The collection spanned nouns, adjectives, verbs, and adverbs and integrated insights from various emotion representation theories. For example, the adjective *cheerful* was semantically linked with the noun *cheerfulness,* the verb *cheer up,* and the adverb *cheerfully.* This approach can target positive, negative, ambiguous, or neutral emotions. For instance, when provided with the term *difficulty,* the system proposes emotions such as *concern* and *apathy.* Similarly, when given the word university along with the specific emotional tone (e.g., positive), the system recommends terms that resonate with positive emotions (such as scholarship or achievement) and are associated with feelings such as *enthusiasm.*

AffectNet (Cambria & Hussain, 2012) comprises 10 thousand words aligned with WORDNET-AFFECT. This tool broadens the WORDNET-AFFECT set of labels to cover concepts such as have breakfast. Its so-called Fuzzy Affect Lexicon has around 4 thousand entries, each manually annotated by a linguist with one out of 80 possible emotion labels.

The NRC-Emotion Lexicon (Mohammad & Turney, 2013) contains around 14 thousand English terms associated with the eight fundamental emotions proposed by Plutchik (1980) -*anger, fear, anticipation, trust, surprise, sadness, joy, and disgust*- plus positive and negative sentiment. The Lexicon terms were associated with each emotion via crowd-sourcing manual annotations on Mechanical Turk. The annotators assigned a score of 1 when the word was associated with an emotion or sentiment (positive vs negative) and zero when it was not. The lexicon includes automatic translations of each word in over 100 languages.

The NRC Hashtag Emotion Lexicon (Mohammad & Kiritchenko, 2013) is one of the first emotion lexicons employing a data-driven approach. Its 16,862 unigrams (words) were automatically associated with Plutchik's eight fundamental emotions. The words are taken from tweets containing emotion-related hashtags, such as #happy or #anger.

EmoSenticNet (Poria et al., 2013) is a lexicon created to bridge sentiment analysis and emotion recognition. The lexicon consists of strings of text tagged with both sentiment (positive or negative) and WORDNET-AFFECT emotion labels. It is an extension of WORDNET-AFFECT's emotion labels and encompass a broader set of vocabulary.

The DepecheMood lexicon (Staiano & Guerini, 2014) has 37,000 terms taken from news articles from the social news network *rappler.com*. The words were ingeniously annotated through crowdsourcing. The *rappler*'s readers can react to news articles using a set of mood meters or emotion-based labels. These mood meters let the readers/annotators expressing their feelings by selecting from a range of emotions (afraid, amused, happy, sad, angry, and annoyed, inspired and don't care).

The Linguistic Inquiry and Word Count (LIWC, Pennebaker et al., 2015) consists of approximately 1,400 English terms manually selected to represent affect categories such as *anxiety, anger,* and *sadness.*

3.3 Annotated Datasets for the Detection of Emotion Categories

The annotated datasets for emotion detection provide a valuable foundation to analyse and detect emotion categories within their textual context. In the following, we report a list and a brief description of the dataset annotated for emotion categories:

ISEAR (Scherer & Wallbott, 1994) is managed by the Swiss Centre for Affective Sciences. This dataset stems from the International Survey on Emotion Antecedents and Reactions (ISEAR) and includes responses from 3,000 participants across 37 countries and has 7,665 sentences labelled with emotion categories. The respondents reported situations in which they experienced 7 emotions (*joy, fear, anger, sadness, disgust, shame,* and *guilt*), how they appraised the situations, and how they reacted.

AMAN'S Emotion dataset (Aman & Szpakowicz, 2007) contains 1,466 sentences from blog posts labelled with 6 emotion categories (*happiness, sadness, disgust, anger, fear, surprise*), mixed emotion (two or more emotions expressed simultaneously) and no emotion.

Emotion-Stimulus data (Ghazi et al., 2015) is a dataset of 2,414 sentences with emotions and their causative stimuli annotated. The sentences are annotated following Ekman's 6 basic emotions, plus *shame.*

Smile dataset (Wang et al., 2016) consists of 3,085 tweets from British Museum-affiliated Twitter handles, annotated for five emotions (*anger, disgust, happiness, surprise,* and *sadness*).

WASSA-2017 Emotion Intensities (EmoInt) Data (Mohammad & Bravo-Marquez, 2017) is designed for detecting tweet emotion intensity. It is labelled for 4 discrete emotions (*anger, fear, joy,* and *sadness*).

Daily Dialog (Li et al., 2017) comprises human-written dialogical conversations for a total of 13,118 sentences labelled for Ekman's 6 discrete emotions, plus the neutral state.

Grounded Emotion data (Liu et al., 2017) was collected to explore the influence of external factors on tweeters' emotions. It comprises 2,557 tweets annotated with the emotions *happy* or *sad*.

MELD dataset (Poria et al., 2017) is a multimodal set with data from Friends TV Show. It contains over 14,000 dialogues and utterances labelled for *anger, disgust, sadness, joy, surprise, fear,* and the neutral state.

Emotion lines (Chen et al., 2018) consists of data from the show Friends and Facebook Messenger chats, totalling over 29,000 utterances that are labelled with Ekman's 6 basic categorical emotions, plus neutral.

Affect Detection in Tweets- SemEval-2018 Task 1 (Kravchenko & Pivovarova, 2018) the authors tested the performance on the manually annotated dataset on the affect detection task. The task identified which features generate the most informative sentence representations. The results show that word embeddings provide more informative sentence representations than lexicon features alone. However, combining lexicon features with embeddings leads to higher performance than using embeddings alone.

EMOEvent (Plaza del Arco et al., 2020) is a multilingual emotion dataset centred around various events from April 2019. They gathered tweets from Twitter, and three Amazon Turk annotators assigned to each tweet one of seven emotions. These labels included Ekman's six fundamental emotions, plus neutral, and the category *other emotions* for unclear cases. The dataset has 8,409 tweets in Spanish and 7,303 in English. Furthermore, every tweet was categorised as either offensive or non-offensive.

3.4   Lexicons for the Detection of Emotion Dimensions

Most lexicons based on the dimensional model of emotions annotate text valence (positive or negative emotions). A very small number of lexicons annotate other emotion dimensions, such as arousal and dominance. Lexicons with emotion valence and intensity are particularly useful for tasks requiring a nuanced interpretation of emotions, such as sentiment analysis and content recommendation.

In the following, we describe the most prominent lexicons for emotion dimensions:

ANEW (Bradley & Lang, 1999) is a standout in NLP emotion detection, as emotion psychologists Bradley and Lang created it. ANEW lexicon has about 2,000 words, each rated on three emotion dimensions: valence, arousal, and dominance.

AFINN (Nielsen, 2011) is one of the earliest and most straightforward emotion dimension lexicons. AFINN presents words scored for the intensity of valence, ranging from very negative (−5 valence score) to very positive (+5 valence score).

NRC Hashtag Sentiment Lexicon (Mohammad et al., 2013) has 679,468 entries, focusing on negative and positive emotion valence. It automatically categorises words of Twitter hashtags.

Warriner et al.,'s 2013 **Lexicon** expands on the ANEW lexicon to encompass close to 14,000 English lemmas. As in ANEW, this lexicon contains ratings of valence, arousal, and dominance. It also highlights the variations among the ratings based on the raters' gender, age, and education.

Sentiment140 Affirmative Context Lexicon and Sentiment140 Negated Context Lexicon (Kiritchenko et al., 2014) contains 330,324 entries in its affirmative context and 43,984 in the negated context. The negated context terms are words in a linguistic context in which the statement or assertion is denied or contradicted. Negation is commonly achieved through words such as *not, never, neither, none,* and *no* among others. The lexicon categorises emotion valence as either negative or positive, and it was automatically generated using hashtags from Twitter.

Reviews Lexicon (Liu et al., 2015) is a manually annotated lexicon stemming from the collection of online product reviews. It provides a comprehensive list of words valence (positive or negative).

The NRC Valence, Arousal, and Dominance (VAD) Lexicon (Mohammad, 2018) comprises over 20,000 English words, each assigned valence, arousal, and dominance scores. The scores range from 0 (for Valence, Arousal and Dominance) to 1 (for Valence, Arousal and Dominance), and the scores are manually annotated. This lexicon is significantly more extensive than any previous dimension lexicon. Additionally, Mohammed demonstrated that the ratings in this lexicon are considerably more reliable than any other dimension lexicon.

3.5   Annotated Datasets for the Detection of Emotion Dimensions

The annotation of emotion dimensions can help focus on the emotion intensity and its nuances. For example, *content* and *ecstatic* are both positive emotions, but they vary in terms of arousal. Such granularity enables more accurate emotion detection and sentiment analysis. For applications like virtual assistants or chatbots, understanding the nuances of human emotion can lead to more contextually appropriate responses and an improvement in the user experience. In recommendation systems, capturing emotion dimensions can lead to more personalized content suggestions. Such annotations can lead to a richer insight into the data, beneficial across fields like marketing, social sciences, and public health. In the following, we describe annotated datasets designed specifically for detecting emotion dimensions in text.

EmoTales (Francisco et al., 2012) is a narrative-focused corpus comprising 1,389 English sentences from 18 folk tales, annotated by 36 raters. It represents emotions through both emotional categories and dimensions. For emotion categories, 119 distinct emotional labels. Regarding emotional dimensions, EmoTales is rated for evaluation (sentiment), activation (intensity), and power (empowered/powerless). For valence, there are positive vs negative emotions, such as happiness, satisfaction, hope, unhappiness, dissatisfaction, and despair. Activation represents the activity vs passivity scale of emotions, with emotions such as excitation at one extreme and calmness and relaxation at the other. Power represents the sense of control which the emotion wields on a person. At one end of the scale, we have emotions such as in control and calm and on the other, we find anxious, fear and submission.

The valence and arousal of Facebook posts (Preotiuc-Pietro et al., 2016) is a dataset comprising 2895 social media entries, evaluated by two annotators trained in psychology. The authors used two nine-point ordinal scales to assess each post: one scale measured valence (sentiment), while the other gauged arousal (intensity).

EMOBANK (Buechel & Hahn, 2017) includes over 10,000 sentences annotated using the Valence-Arousal-Dominance emotion model. Part of this dataset employs Ekman's six basic emotions. The data sources are diverse, encompassing blogs, fiction, and newspapers.

SemEval 2018 dataset (Mohammad et al., 2018) was the dataset offered at the International Workshop on Semantic Evaluation in 2018 for the task-1 Affect in Tweets. It is a dataset for emotion intensity/valence detection.

Affect Detection in Tweets-SemEval-2018 Task 1 (Gupta & Yang, 2018) the authors presented a system designed to predict emotion intensity in tweets. The system analysed a Twitter message using features derived from parts-of-speech, n-grams, word embeddings (see Chap. 4), and various affective lexicons, including AFFIN, NRC Emotion & Hash Emotion, and the custom-developed EI Lexicons. The results indicate that incorporating affective lexicon features significantly enhanced the prediction performance of the system and uncovered interesting associations at both the word level and message level.

SemEval-2023 Task 12 (Belbachir, 2023) dataset comprises Twitter data in 14 low-resource African languages. The primary objective of the Task was to compare the performance of three sentiment analysis models across these languages.

3.6   Conclusion

This chapter explored some of the most used resources for emotion detection in English, along with a few examples in other less-resourced languages. Lexicons provide a straightforward way to map words to emotions, making them effective for classification. Annotated datasets, on the other hand, offer detailed contextual information, which aids in developing sophisticated models to capture the complexities of emotional expression. Understanding these resources is crucial as we move forward to examine the evolution from rule-based methods to machine learning and deep learning techniques for emotion detection. With a solid grasp of these critical resources, we are better prepared to understand the historical context and development of various emotion detection techniques. Chapter 4 will focus on the initial methods, which relied on keyword analysis and rule-based systems, laying the foundation for more advanced machine-learning approaches.

References

Aman, S., & Szpakowicz, S. (2007). Identifying expressions of emotion in text. In V. Matoušek, & P. Mautner (Eds.) *Text, Speech and Dialogue. TSD 2007*. Lecture Notes in Computer Science (4629), Springer.

Artstein, R., & Poesio, M. (2008). Inter-coder agreement for computational linguistics. *Computational Linguistics, 34*(4), 555–596.

Belbachir, F. (2023). Foul at SemEval-2023 task 12: MARBERT language model and lexical filtering for sentiments analysis of tweets in Algerian Arabic. In *Proceedings of the 17th International Workshop on Semantic Evaluation (SemEval-2023)* (pp. 389–396).

Bradley, M.M., & Lang, P.J. (1999). *Affective norms for English words (ANEW): Instruction manual and affective ratings.* https://pdodds.w3.uvm.edu/teaching/courses/2009-08UVM-300/docs/others/everything/bradley1999a.pdf

Buechel, S., & Hahn, U. (2017). EmoBank: Studying the impact of annotation perspective and representation format on dimensional emotion analysis. In *Proceedings of the 15th Conference of the EUROPEAN CHAPTER of the Association for Computational Linguistics* (Vol. 2, pp. 578–585).

Cambria, E., & Hussain, A. (2012). *Sentic computing.* Springer.

Chen, S-Y., Hsu, C-C., Kuo, C-C., Ku, L-W., et al. (2018). *Emotionlines: An emotion corpus of multiparty conversations.* arXiv 1802.08379.

Ekman, P., & Friesen, W. V. (1969). The repertoire or nonverbal behavior: Categories, origins, usage and coding. *Semiotica, 1*, 49–98.

Francisco, V., Hervás, R., Peinado, F., & Gervás, P. (2012). EmoTales: Creating a corpus of folk tales with emotional annotations. *Language Resources and Evaluation, 46*, 341–381.

Ghazi, D., Inkpen, D., & Szpakowicz, S. (2015). Detecting emotion stimuli in emotion-bearing sentences. In A. Gelbukh (Ed.) *Computational linguistics and intelligent text processing*. Lecture Notes in Computer Science, 9042. Springer.

Gupta, R. K., & Yang, Y. (2018). CrystalFeel at SemEval-2018 task 1: Understanding and detecting emotion intensity using affective lexicons. In *Proceedings of the 12th International Workshop on Semantic Evaluation* (pp. 256–263). Association for Computational Linguistics.

Kiritchenko, S., Zhu, X., Cherry, C., & Mohammad, S. (2014). NRC-Canada-2014: Detecting aspects and sentiment in customer reviews. In *Proceedings of the 8th International Workshop on Semantic Evaluation* (pp. 437–442). Association for Computational Linguistics.

Kravchenko, D., & Pivovarova, L. (2018). DL Team at SemEval-2018 task 1: Tweet affect detection using sentiment lexicons and embeddings. In *Proceedings of the 12th International Workshop on Semantic Evaluation* (pp. 172–176).

Li, Y., Su, H., Shen, X., Li, W., Cao, Z., & Niu, S. (2017). DailyDialog: A manually labelled multi-turn dialogue dataset. In *Proceedings of the Eighth International Joint Conference on Natural Language Processing* (Vol. 1, pp. 986–995). Asian Federation of Natural Language Processing.

Liu, P., Joty, S., & Meng, H. (2015). Fine-grained opinion mining with recurrent neural networks and word embeddings. In *Proceedings of the 2015 Conference on Empirical Methods in Natural Language Processing* (pp. 1433–1443).

Liu, V., Banea, C., & Mihalcea, R. (2017). Grounded emotions. *Seventh international conference on affective computing and intelligent interaction, 2017*, 477–483.

Mehrabian, A., & Russell, J. A. (1974). *An approach to environmental psychology.* MIT Press.

Miller, G. A., Beckwith, R., Fellbaum, C., Gross, D., & Miller, K. J. (1990). Introduction to Word-Net: An on-line lexical database. *International Journal of Lexicography, 3*(4), 235–244.

Mohammad, S., Kiritchenko, S., & Zhu, X. (2013). NRC-Canada: Building the state-of-the-art in sentiment analysis of tweets. In *Second Joint Conference on Lexical and Computational Semantics, Vol. 2, and Proceedings of the Seventh International Workshop on Semantic Evaluation* (pp. 321–327).

Mohammad, S., & Bravo-Marquez, F. (2017). Emotion intensities in tweets. In *Proceedings of the 6th Joint Conference on Lexical and Computational Semantics* (pp. 65–77).

Mohammad, S., Bravo-Marquez, F., Salameh, M., & Kiritchenko, S. (2018). SemEval-2018 task 1: Affect in tweets. *Proceedings of the 12th International Workshop on Semantic Evaluation*, 1–17.

Mohammad, S. (2018). Obtaining reliable human ratings of valence, arousal, and dominance for 20,000 English words. In *Proceedings of the 56th Annual Meeting of the Association for Computational Linguistics* (Vol. 1, Long Papers, pp. 174–184).

Mohammad, M., & Turney, P. D. (2013). Crowdsourcing a word–emotion association lexicon. *Computational Intelligence, 29*(3), 436–465.

Mohammad, S., & Kiritchenko, S. (2013). Using nuances of emotion to identify personality. *Proceedings of the International AAAI Conference on Web and Social Media, 7*(2), 27–30.

Nielsen, F. A. (2011). *A new ANEW: Evaluation of a word list for sentiment analysis in microblogs.* ArXiv:1103.2903.

Ortony, A., Clore, G. L., & Collins, A. (1987). *The cognitive structure of emotions.* Cambridge University Press.

Pennebaker, J.W., Booth, R.J., Boyd, R.L., & Francis, M.E. (2015). *Linguistic Inquiry and Word Count. LIWC2015.* Pennebaker Conglomerates.

Plaza del Arco, F. M., Strapparava, C., Urena Lopez, L. A., & Martin, M. (2020). EmoEvent: A multilingual emotion corpus based on different events. In *Proceedings of the Twelfth Language Resources and Evaluation Conference* (pp. 1492–1498).

Plutchik, R. (1980). A general psychoevolutionary theory of emotion. In R. Plutchik & H. Kellerman (Eds.), *Theories of Emotion* (pp. 3–33). Academic Press.

Poria, S., Gelbukh, A., Hussain, A., Das, D., & Bandyopadhyay, S. (2013). EmoSenticNet. https://www.researchgate.net/publication/235971757_emosenticnet

Poria, S., Hazarika, D., Majumder, N., Naik, G., Cambria, E., & Mihalcea, R. (2017). *MELD: A multimodal multi-party dataset for emotion recognition in conversations.* ArXiv:1810.02508.

Preoţiuc-Pietro, D., Schwartz, H. A., Park, G., Eichstaedt, J., Kern, M., Ungar, L., & Shulman, E. (2016). Modelling valence and arousal in Facebook posts. In *Proceedings of the 7th Workshop on Computational Approaches to Subjectivity, Sentiment and Social Media Analysis* (pp. 9–15).

Scherer, K. R., & Wallbott, H. G. (1994). Evidence for universality and cultural variation of differential emotion response patterning. *Journal of Personality and Social Psychology, 66*, 310–328.

Staiano, J., & Guerini, M. (2014). Depeche Mood: a lexicon for emotion analysis from crowd annotated news. In *Proceedings of the 52nd Annual Meeting of the Association for Computational Linguistics* (Vol. 2, 427–433).

Strapparava, C., Valitutti, A., & Stock, O. (2006). The affective weight of lexicon. In *Proceedings of the Fifth International Conference on Language Resources and Evaluation* (pp. 423–426).

Valitutti, A., Strapparava, C., & Stock, O. (2004). Developing affective lexical resources. *Psychology Journal, 2*(1), 61–83.

Wang, B., Liakata, M., Zubiaga, A., Procter, R. & Jensen, E. (2016). SMILE: Twitter emotion classification using domain adaptation. In *Workshop on Sentiment Analysis Where AI Meets Psychology* (pp. 15–21).

Warriner, B., Kuperman, V., & Brysbaert, M. (2013). Norms of valence, arousal, and dominance for 13,915 English lemmas. *Behavior Research Methods, 45*(4), 1191–1207.

Rule-Based Systems for Emotion Detection 4

4.1 Foundations of Rule-Based Emotion Detection Systems

In the early stages of emotion detection studies, researchers heavily relied on rule-based systems, which use predefined rules and patterns derived from specialised lexicons and datasets. These systems, manually crafted to identify keywords, phrases, and syntactic structures associated with a variety of emotions, offer a structured and interpretable approach to emotion detection. Moreover, the deterministic nature of rule-based models allows for integrating linguistic and psychological insights, making them a practical solution for early emotion detection tasks. The early rule-based systems laid the groundwork for more sophisticated emotion detection methods. They essentially demonstrated the feasibility of automated text analysis for emotion and sentiment analysis, emphasising the importance of domain expertise in creating effective models.

Initially, rule-based systems focused on sentiment analysis, aiming to determine the sentiment expressed in a text—typically categorised as positive, negative, or neutral. As many companies underwent digital transformation, the challenge was converting the surge of unstructured written text into actionable insights for decision-making. Rule-based systems played a crucial role in understanding public opinion and sentiment in product reviews, customer feedback, and social media posts. These systems employed lexicons listing sentiment-laden words and phrases and their associated polarity. Rule-based models could assign an overall sentiment score to the text by analysing the presence and frequency of these sentiment indicators.

Turney (2002), together with Tong (2001) and Hatzivassiloglou and McKeown (1997), were among the first authors to explore sentiment by classifying the semantic direction (positive or negative) of adjectives. However, as suggested by Ortony et al. (1988), it is crucial to differentiate between words that directly describe emotions, such as *fear* or *cheerful*, and those that imply emotional meanings based on context, such as *monster* or

© The Author(s), under exclusive license to Springer Nature Switzerland AG 2025
F. Cavicchio, *Emotion Detection in Natural Language Processing*, Synthesis Lectures on Human Language Technologies, https://doi.org/10.1007/978-3-031-72047-5_4

cry. The former are labelled as direct affective words, while the latter are termed indirect affective words.

As we have seen in Chap. 3, to effectively detect emotions in a text, it is necessary to build lexicons and annotated datasets of direct and indirect affective words. Rule-based approaches to emotion detection depend on predefined rules and patterns, typically developed using specialised lexicons and annotated datasets. Lexicons are essentially dictionaries linking words or phrases to specific emotion categories. A rule-based system employing a lexicon classifies text by identifying specific keywords or phrases found within the lexicon. In contrast, annotated datasets are extensive collections of texts annotated with emotion labels, providing rich contextual examples for different emotion categories. These annotated datasets enable the creation of more complex rules that account for sentence structure and surrounding text.

4.2   Rule-Based Methods for Emotion Detection

The first stage in detecting emotions from text involves preparing the data for the rule-based or keyword-based detection of emotion models. Rule-based models rely on predefined rules and patterns, often created using data from specialised lexicons and annotated datasets. These rules classify or predict outcomes based on the presence of specific keywords, phrases, or patterns in the data. In emotion detection, for example, a rule-based model might classify a text as *happy* if it contains keywords associated with *happiness*. While rule-based approaches do not always involve numerical data, they can include numerical thresholds or conditions. For example, if a rule sets a threshold of three keywords associated with *joy* to classify the text as *joyful*, a text containing *happy* more than three times is classified as expressing *joy*. Similarly, a rule might classify a text as expressing sadness if the keyword *sad* appears more than twice.

However, consider the text: *I am so happy today because everything went well, but I also felt a bit bitter thinking about the people we had to leave behind*. A rule with a numerical threshold of three joy-related terms would identify *happy* and *well*, which appear twice, while the rule for *sadness* would recognize *bitter*. Although neither emotion meets its numerical threshold, the model classifies the dominant emotion as *joy* due to the higher frequency of joy-related keywords. In this example, *bitter* is a stronger negative word than *well* is a positive word, but a rule-based model that relies solely on numerical thresholds does not account for the intensity of word sentiment, leading to the text being considered more positive than negative. Figure 4.1 presents the flowchart demonstrating how the rule-based model discussed above processes the sentence, *I am so happy today because everything went well, but I also felt a bit bitter thinking about the people we had to leave behind*. The model identifies emotion-related keywords, checks them against predefined numerical thresholds, and ultimately determines the dominant emotion based on keyword frequency.

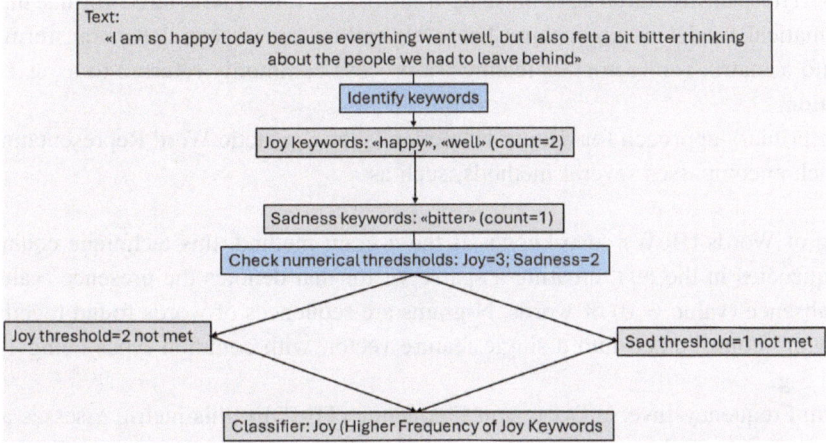

Fig. 4.1 Flowchart illustrating the analysis of the sentence *I am so happy today because every-thing went well, but I also felt a bit bitter thinking about the people we had to leave behind*. The flowchart includes separate paths for identifying and evaluating *joy* and *sadness* keywords against their respective numerical thresholds. It concludes with the classification of the dominant emotion based on keyword frequency

Text preprocessing and extracting features are crucial steps. In this section we outline methods for cleaning text, tokenizing it, normalization, and enabling the extraction of informative features. We note that, especially in the context of social media and customer or product reviews, which often feature brief text, emojis, slang, and incomplete words, preprocessing the text is essential.

The critical steps in text preprocessing include:

- **Data Cleaning:** this step covers converting all text to lowercase, removing punctuation and digits, eliminating stop words, stripping away hashtags and HTML tags, and expanding contractions.
- **Tokenization/Segmentation:** it involves breaking down the text or strings into individual tokens, each representing a word.
- **Normalization:** it includes stemming and lemmatization, processes that reduce words to their root form.

After we cleaned the text of punctuation, stop words, and tokenised it, we can check each word or phrase for emotion, comparing them with the lexicon. We can aggregate individual emotions from the lexicon to discern the text's overall emotional tone. There are several aggregation methods, from simple to weighted sums considering word placement. To preserve context, bigrams or trigrams can be examined as well. The lexicon-based results are then validated against a tagged dataset, leading to iterative refinement of the results. Thus, unstructured text must be transformed into a structured format through a mathematical model or algorithm. These algorithms are essential for transforming the text into a matrix (or vector) of features, a process commonly referred to as text feature extraction.

The primary approach for feature extraction is the Syntactic Word Representation. This approach encompasses several methods, such as:

- Bag of Words (BoW): also known as the n-gram method, this technique counts word frequencies in the text, creating a sparse vector that denotes the presence (value $= 1$) or absence (value $= 0$) of words. N-grams are sequences of words found together, and they are consolidated into a single feature vector, with common types being 1-g, 2-g, and 3-g.
- Term Frequency-Inverse Document Frequency (TF-IDF): this metric assesses a term's significance or relevance in a document by calculating its frequency in the document against its rarity across a larger annotated dataset. TF is the number of times a term appears in a document/total number of terms in the document, and IDF is the number of documents/number of documents containing the term. Thus, TF measures the frequency of a term, and IDF reflects its importance in an annotated dataset or collection of text.
- Part of Speech Tagging (PoS): this process involves categorising words into their grammatical function, such as nouns, verbs, adjectives, and adverbs.
- Named Entity Recognition (NER): it involves identifying and classifying entities within a sentence, such as individuals, brands, etc., and categorising them into groups, such as dates, organisations, persons, times, locations, etc.

The predominant rule-based methods for detecting emotions in text include keyword and rule-based methods:

- **Keyword-based approach:** this method relies on identifying specific keywords in a text and comparing them with annotations in an annotated dataset. Initially, a list of emotion-related keywords is created using lexicons or annotated datasets. Following this, the dataset undergoes pre-processing. The keyword approach then involves matching keywords from the text with those in a predefined list, assessing the strength of these emotion keywords, checking for any negation cues. Studies by Huang and Pao (2012), Ma et al. (2005), Perikos and Hatzilygeroudis (2016), and Shivhare and Khethawat (2015) employ the keyword-based method to detect emotions in text.
- **Rule-based approach:** this technique applies linguistic rules to discern emotions from the text. After cleaning and preparing the text, each word is associated with a probabilistic matrix or vector value for a specific emotion label or dimension. Afterwards,

the most effective rules are selected and used on the test dataset to classify the emotions. This approach is exemplified by Lee et al. (2010), Udochukwu and He (2015), and Liu and Cocea (2017).

4.3   Advantages and Limitations of Rule-Based Approaches

Navigating the vast amount of informal text data generated daily reveals the complexities of detecting emotions in social media. Detecting emotions in texts present multifaceted challenges, not only because of the intricacies of human emotions but also because of the nature of online language—typos, slang, and grammatical inconsistencies—complicating the automated analysis processes. Rule-based and keyword-based methods stand out for their versatility. They rely on predefined dictionaries or lists of words associated with specific sentiments or emotions.

Methods based on annotated datasets are developed by annotating vast amounts of text data and are usually domain specific. Through the annotated data, we can then extract rules and linguistic patterns. The annotated dataset specificity permits higher precision when we detect emotion categories or dimensions (e.g., valence). For instance, an annotated dataset developed for analysing medical reviews might be exceptionally adept at recognising emotion valence in that context. However, the same annotated dataset will likely fail to detect most emotions when applied, for example, to movie reviews. Therefore, while they offer greater accuracy within their designated domains, annotated datasets may be limited when taken outside their context. While they excel in targeted scenarios, they might not be the best choice for generalised or cross-domain applications. Finally, while crafting rules for specific data can be straightforward, the task becomes less and less successful when we deal with an extensive volume of data coming from different sources.

As we have seen in the previous paragraph, one of the most employed rule-based approaches to emotion detection is keyword recognition. This approach entails the creation or utilisation of lexicons and annotated datasets annotated by emotion categories and/or dimensions. In Chap. 3, we have illustrated several well-known lexicons, including WordNet-Affect (Valitutti et al., 2004), DepecheMood (Staiano & Guerini, 2014), and the NRC lexicon (Mohammad & Turney, 2013). As we have seen, these lexicons associate terms with emotion categories such as *happy*, *hate*, *angry*, and *sad*, or dimensions such as *positive*, *negative*, and levels of emotion intensity, which we can consider as our keywords. The goal of a keyword recognition system is to pinpoint the relevant keywords in a new given text at the sentence level. In the case of emotion category detection, when one of the lexicon's keywords aligns with the text's content, the corresponding emotion label is attributed to that sentence. The most straightforward approach is when the emotion lexicon lists the emotion category joy, and the text states, *I was overwhelmed with joy seeing my home after a month away*. In this case, the emotional label joy is applied to that sentence. However, keyword recognition faces several problems. These problems include the lexicon limitations in the number of emotion labels, the potential keyword ambiguity, and the absence of linguistic context. More sophisticated lexicons, such as DepecheMood, do not have a one-to-one match between a word and an emotion. Instead, these lexicons may provide a vector or a score that represents the association strength of a term with multiple emotion categories or dimensions. In the following, we report examples of application of the keyword method.

- **Contextual Meaning:** Advanced lexicons can be used for context disambiguation
  Example: *The weather was sick today.*
- **Synonyms:** Lexicons can be used to identify synonyms not related to basic emotion terms (in the example below, surprise party)
  Example: *She was exuberant at the surprise party.*
- **Intensity Levels:** Lexicons can help capturing the intensity of emotions
  Example: *He was slightly annoyed by the noise.*
- **Negations:** Lexicons can be used to identify sentence negation in the context of emotions
  Example: *I'm not feeling too thrilled about the test.*
- **Multiple Emotions:** Lexicons can be used to identify multiple emotions in a sentence
  Example: *I'm sad about leaving but excited for a new start.*

Another popular method is the lexical affinity method. This method assigns a probability to emotion-related words. For example, the term *good* has a 90% likelihood of being interpreted as positive. However, in the sentence *too much of a good thing can be dangerous*, the presence of *good* does not necessarily imply an entirely positive valence for the overall statement.

Although most lexical affinity systems default to a binary *positive* or *negative* categorisation, some can identify neutral tones or even pinpoint emotion categories such as *joy*, *sadness*, or *anger*. The crucial consideration is, again, context. For example, a word such as *cold* can bear a negative connotation in scenarios like a *cold response*, yet it might lean neutral or even positive when describing *cold ice cream on a hot day*. The role of context highlights the necessity for more advanced analysis, including syntactic or semantic parsing. Additionally, we need to consider word combinations. A classic example is the phrase *not bad*. While *not* and *bad* might be seen as negative, their combined meaning typically is mildly positive.

Another obstacle to detecting emotion in text is the complexity and wealth of emotional expression. Human emotions are layered with different intensity levels, even within a single sentence. This complexity is further accentuated when we use metaphors and context-driven emotional cues. For example, in the sentence *Ugh, this day was soooooo long!* The lengthening in *soooooo* emphasises the length of the day. However, it remains ambiguous whether the speaker found the day tedious, exhausting, or perhaps even fulfilling. In the sentence *Just what I needed on a Monday morning, a spilled coffee,* the statement is also clearly ironic, indicating the speaker's dismay at the unfortunate event. To disambiguate these types of sentences, the lexical affinity method is frequently paired with supplemental techniques such as syntactic parsing, context-sensitive anayses and word embedding. Such integrations aim to show deeper layers of linguistic meaning and bolster the accuracy of emotion detection.

The use of emotion-annotated datasets can enhance the lexical affinity method by fine-tuning the associated probability scores through the co-occurrence of words with emotion-annotated words or phrases. While the internet has abundant textual information, structured and annotated data are crucial for producing interpretable outputs. Manual annotations of emotions must be precise and reproducible, with validation using measures such as the kappa statistic to ensure annotator reliability. Kappa ensures annotation consistency and validates the accuracy and reliability of the emotion categories and dimensions assigned within the dataset. However, a common issue with emotion-annotated datasets is language diversity, as many existing resources are only in English. This limitation needs to be addressed when considering domain-specific lexicons or web slang popularised by platforms such as Reddit. It is important to note that over time, annotated datasets and lexicons have been translated into various languages. Additionally, multilingual annotated datasets have been compiled to address the challenges posed by language diversity.

Rule-based emotion detection models use predefined rules from lexicons and annotated datasets that link specific words or phrases to certain emotion categories and dimensions. However, these systems can struggle with the complexities of language. For instance, the phrase *Great, another Monday morning!* can mislead a system into interpreting *great* as positive, missing the sarcasm. Similarly, words can have different sentiments in different contexts, such as *I'm feeling blue about the clear blue sky*, where *blue* can mean sadness or a colour. Negations such as *I can't say I'm thrilled with the results* and idiomatic expressions such as *He's not the sharpest tool in the shed* present additional challenges. Compound sentences with mixed emotions and modifiers that adjust emotional intensity, such as *I'm somewhat happy with the performance, but it could be better*, can also be complex to interpret.

On one hand, specific lexicons, such as the NRC Emotion Lexicon with built-in negations, can be used to provide more nuanced interpretations of negated phrases. Using curated annotated datasets tailored to a dataset's linguistic nuances can improve the accuracy of rule-based models. On the other hand, meticulous lexicons and custom annotated datasets, while achieving high accuracy, are a manual endeavour requiring substantial time and effort. Moreover, annotated datasets are typically domain- and language-specific, which can hinder the expansion of rule-based models into new linguistic contexts.

The limitations of keyword and rule-based methods became increasingly evident when dealing with large volumes of data automatically scraped from social media platforms such as Twitter. Concurrently, machine learning algorithms improved, becoming faster and more adept at analysing extensive datasets and identifying patterns. In the next Chapter, we will explore how advancements in machine learning algorithms have significantly enhanced emotion detection systems.

References

Hatzivassiloglou, V., & McKeown, K. R. (1997). Predicting the semantic orientation of adjectives. In *35th Annual Meeting of the Association for Computational Linguistics and 8th Conference of the European Chapter of the Association for Computational Linguistics* (pp. 174–181).

Huang, W-Y., & Pao, T-L. (2012). A study on the combination of emotion keywords to improve the negative emotion recognition accuracy. in *6th International Conference on New Trends in Information Science, Service Science and Data Mining* (ISSDM2012) (pp. 499–503).

Lee, S. Y. M., Chen, Y., & Huang. C. R. (2010). A text-driven rule-based system for emotion cause detection. In *Proceedings of the NAACLHLT 2010 Workshop on Computational Approaches to Analysis and Generation of Emotion in Text* (pp. 45–53).

Liu, H., & Cocea, M. (2017). Fuzzy rule-based systems for interpretable sentiment analysis. In *Ninth International Conference on Advanced Computational Intelligence* (pp. 129–136).

Ma, C., Prendinger, H., & Ishizuka, M. (2005). Emotion estimation and reasoning based on affective textual interaction. In J. Tao, T. Tieniu, & R. W. Picard (Eds.), *Affective Computing and Intelligent Interaction* (pp. 622–628). Springer.

Mohammad, M., & Turney, P. D. (2013). Crowdsourcing a word–emotion association lexicon. *Computational Intelligence, 29*(3), 436–465.

Ortony, A., Clore, G. L., & Collins, A. (1988). *The cognitive structure of emotions.* Cambridge University Press.

Perikos, I., & Hatzilygeroudis, I. (2016). Recognizing emotions in text using ensemble of classifiers. *Engineering Applications of Artificial Intelligence, 51*, 191–201.

Shivhare, N. S., & Khethawat, S. (2015). *Emotion detection from text.* ArXiv:1205.4944.

Staiano, J., & Guerini, M. (2014). Depeche Mood: a lexicon for emotion analysis from crowd annotated news. In *Proceedings of the 52nd annual meeting of the association for computational linguistics* (Vol. 2, pp. 427–433).

Tong, R.M. (2001) An operational system for detecting and tracking opinions in on-line discussion. In *Proceedings of SIGIR Workshop on Operational Text Classification.*

Turney, P. (2002). Thumbs up or thumbs down? Semantic orientation applied to unsupervised classification of reviews. In *Proceedings of the Association for Computational Linguistics* (pp. 417–424).

Udochukwu, O., & He, Y. (2015). A rule-based approach to implicit emotion detection in text. In C. Biemann, S. Handschuh, A. Freitas, F. Meziane, & E. Métais (Eds.) In *Natural Language Processing and Information Systems.* Lecture Notes in Computer Science, Vol. 9103. Springer.

Valitutti, A., Strapparava, C., & Stock, O. (2004). Developing affective lexical resources. *Psychology Journal, 2*(1), 61–83.

Machine Learning Approaches to Emotion Detection

5

5.1 Unsupervised and Supervised Methods for Emotion Detection

Interest in emotion detection in texts has surged due to several factors, including advancements in data analytics methods, the increased availability of lexicons and annotated datasets for emotions, and the notable rise in social media participation. This influx of extensive user data has created a vast repository of user opinions, reactions, and emotions. However, extracting meaningful insights and analysing emotions from such unstructured data has become an ever-increasing challenge. In Chap. 4 we have discussed rule-based methods, which rely on keywords and manually predefined rules, and showed that while they offer quick adaptations and are straightforward to implement, they often struggle with large quantity of data and rule generalization. In this Chapter, we will discuss the shift from rule-based approaches to machine-learning methods for detecting emotions. We will present word-embedding, unsupervised, and supervised methods for classifying emotions, highlighting how each method enhances the accuracy of emotion detection in textual data. We start with Fig. 5.1, which visually compares the modules used in rule-based and machine learning-based emotion detection techniques.

Rule-based models operate through a sequence starting with a dataset, followed by text preprocessing, rule extraction, rule selection, and finally, applying these rules to label the emotion. This model relies on explicit rules, often tied to specific keywords or patterns associated with emotions, and its effectiveness is contingent on the precision and comprehensiveness of these rules.

In contrast, the machine learning-based model began with raw data and labelled emotions (supervised learning). The process involves text preprocessing and feature extraction and then uses these features to train a machine learning algorithm. The trained model can then predict emotions in new texts and label them accordingly. This model adapts and

© The Author(s), under exclusive license to Springer Nature Switzerland AG 2025
F. Cavicchio, *Emotion Detection in Natural Language Processing*, Synthesis Lectures
on Human Language Technologies, https://doi.org/10.1007/978-3-031-72047-5_5

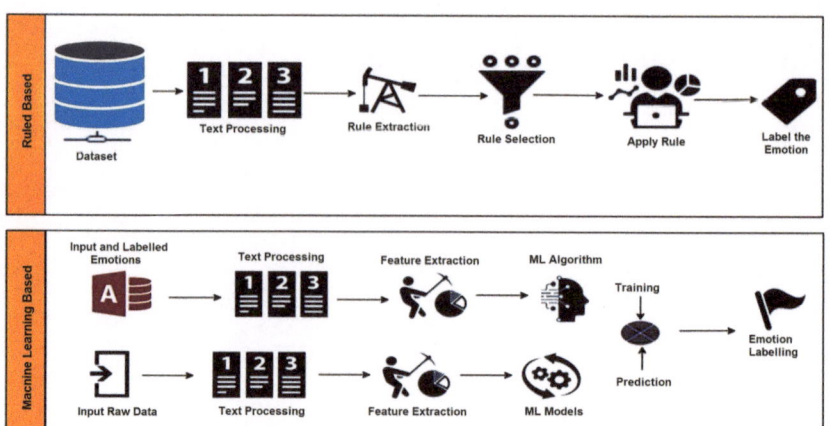

Fig. 5.1 A comparison between rule-based and machine learning modules for the detection of emotion labels from textual data (adapted from Kusal et al., 2021, p. 9)

improves over time, learning from the dataset and capable of handling complex variations in text.

In summary, while the rule-based model is straightforward and rule-dependent, the machine learning model, learning from data, can effectively capture nuanced emotional expressions, making it better suited for applications requiring scalability and adaptability.

In the context of machine learning, word embeddings are a powerful tool that converts words into numerical vectors, capturing semantic meanings based on word usage in large corpora. Thus, word embeddings enable machine learning algorithms to effectively learn and predict emotional states from text, making the system more sensitive to the subtleties of human language. The next paragraph will explore word embeddings in greater detail, discussing their significant impact on improving emotion detection within machine learning models.

5.1.1 Word Embeddings and Emotions

Word embeddings can help capture multiple dimensions of meaning by situating words in high-dimensional spaces, thus allowing for emotion detection based on the context. Unlike the Bag-of-Words (BoW) approach, which treats text as a collection of unordered words, word embeddings account for the context and semantic relationships between words, making them computationally efficient and linguistically rich for the emotion detection task.

Word embeddings represent words as vectors in a high-dimensional space, where similar meanings have similar representations. They are beneficial for an array of NLP tasks, including machine translation, word analogy, part of speech tagging, sequence labelling, named entity recognition, text classification, and speech processing. The foundational concept of word embedding techniques is rooted in the distributional hypothesis, famously summarised as *You shall know a word by the company it keeps* (Firth, 1957). The validity of distributional principles has been supported by numerous psycholinguistics studies (e.g., Charles, 2000). Word embedding techniques use statistical data, such as the frequency with which words appear together, to embed linguistic rules and patterns within vectors. The result is a vector space in which a vector represents each distinct word in a corpus. Similar words are positioned closely within this space.

The prevalent word embedding models are Word2Vec and GloVe. Word2Vec and GloVe transform the text into vectors by learning context through word co-occurrences. Word2Vec (Mikolov et al., 2013) is a predictive approach to language modelling trained by predicting adjacent words around a target word or the reverse. Its structure relies on a neural network and is available in two variants: Continuous Bag of Words (CBOW) and Skip-gram. CBOW predicts a word based on its context, whereas Skip-gram predicts the context for a given word. Through training the neural network with large corpora, Word2Vec computes the word embeddings for each word. GloVe (Pennington et al., 2014) is a count-based word embedding model, and it operates on the principle of dimensionality reduction on a co-occurrence matrix, which tracks the frequency of word pairs across different contexts. The matrix is then broken down into a smaller matrix of the embedding features. However, it should be noted that, unlike contextual embedding models (e.g., BERT), Word2Vec and GloVe do not directly account for the position and order of words within the sentences. Despite their strengths, standard word embeddings often fail to discern subtle differences in emotions. Emotions often require a more granular approach than word proximity. For instance, *joy* and *contentment* may be found in similar contexts. However, these two emotions convey subtly different levels of positive emotion, which their unique contextual associations can reflect. For example, *joy* might co-occur more frequently with words such as *excited*, *thrilling*, and *spectacular*, whereas *contentment* might be found alongside *peaceful* or *serene*. These contextual differences are crucial for understanding the distinct emotional tones they convey. While standard word embeddings capture general semantic similarity, they often need additional contextual modelling to discern these finer distinctions. Thus, to enhance word embeddings for emotion detection, a solution is adding prior knowledge sourced from lexicons and annotated datasets. This additional data, not captured by word co-occurrence frequencies, can significantly improve word embeddings for emotion detection.

Consequently, recent advancements have introduced sentiment and emotion-aware word embeddings by integrating prior knowledge from emotion lexicons. For example, the model proposed by Tang et al. (2016) introduced an approach known as sentiment embedding, which effectively utilises word contexts alongside sentiment polarity (positive or negative emotions) conveyed in texts to create more dynamic and robust word representations. Giatsoglou et al. (2017) introduced an embedding model merging lexicon-based and word embedding-based data into a unified vector form. Their lexicon data was sourced from EmoLex (Mohammad & Turncy, 2013), covering eight emotions as per Plutchik (1980) with a binary scoring system. Agrawal et al. (2018) used WordNet-Affect (Valitutti et al., 2004) and EmoLex to enrich their word embedding vectors. Wang et al. (2020) developed an emotional embedding model for Chinese, enhancing conventional word embeddings with sentiment and emotion categories lexicons. Khosla et al. (2018) introduced an affect-enriched word representation model that combined word embeddings with affective data based on the Valence-Arousal-Dominance (VAD) model of emotion dimensions.

In conclusion, while word embeddings are computationally efficient and linguistically rich text representations, their application in emotion detection requires additional layers to capture the complex and multi-dimensional nature of human emotions. By integrating emotion-specific knowledge and lexical resources, recent advancements in emotion-aware word embeddings provide more fine-grained and effective tools to analyse emotions in texts.

5.1.2 Unsupervised Methods for the Detection of Emotion Categories

Supervised and unsupervised learning methods are commonly applied for automatically identifying emotional expressions in text. Supervised learning comes with the drawback of needing extensive annotated datasets for training, leading to a time-consuming and costly annotation process. On the other hand, unsupervised machine learning operates on unlabelled data and does not require explicit instructions to learn from the data. Unlike supervised learning, where models are trained on data labelled with the correct answers, unsupervised learning algorithms discover hidden patterns and relationships in raw, unorganised data. This approach is beneficial when the data is abundant, but the labelling is impractical or too costly.

Notable differences are evident between supervised and unsupervised techniques and among various emotion models, impacting diverse areas significantly. In the context of emotions, Kort et al. (2001) suggested a model merging emotion category onto a valence-arousal plane. This hybrid approach has also been employed in other tasks, such as in blog posts where Aman and Szpakowicz (2007) explored the identification of emotion categories and their intensities. Despite this, many researchers have predominantly focused on and assessed supervised methods, usually based on the categorical emotion model.

In unsupervised machine learning, the algorithms identify inherent structures within the dataset. To identify the data structure, we usually use methods such as clustering, where data points are grouped based on shared characteristics, or dimensionality reduction, simplifying complex data to make it more understandable. When applied to emotion detection, unsupervised machine learning would analyse text data without predefined categories, looking for patterns and clusters corresponding to different emotions. For example, we expect that sentences with similar word use will be grouped together, thus identifying words corresponding to a specific emotion. Such an approach can be instrumental in emotion detection because it can uncover subtle and complex emotional states that we might find difficult to categorise into predefined labels. Unsupervised learning can adapt to the nuances and variability of human emotion, potentially leading to more accurate and flexible emotion detection systems.

The Vector Space Model (VSM) is an essential technique in NLP, representing text as vectors in a multidimensional space. Each dimension typically corresponds to a term from the document, with the vector values indicating the term's frequency in the document calculated using the TF-IDF method. As we have seen in Chap. 4, this approach can adjust for multiple text documents because it measures the frequency of a word in a document and adjusts for the word's frequency across multiple documents. One of the critical features of VSM is the use of similarity measures, such as cosine similarity, to determine how closely related two documents are. This measure calculates the cosine of the angle between two document vectors, with values closer to 1 indicating high similarity. Within these frameworks, we can establish cosine similarities by assigning the label *neutral* to an input sentence if its cosine similarity does not surpass a specific threshold, indicating the absence of emotion. If it exceeds this threshold, the sentence is labelled with the emotion corresponding to the emotional vector with the highest similarity. However, due to its high-dimensional nature, VSM often needs dimensionality reduction techniques such as Principal Component Analysis (PCA) or Latent Semantic Analysis (LSA). Indeed, many studies using VSM for emotion categories and/or dimensions in texts usually have an LSA to streamline the number of dimensions. LSA's core concept involves transforming terms or documents into a lower-dimensional vector space, known as the latent semantic space. Studies on emotion detection in texts usually integrate LSA with probabilistic theories such as Bayes' rules, facilitating the discovery of latent topics, the associations between documents and topics, and the relationships between words and topics.

For example, Kim et al. (2010) used six lists of emotional words derived from WordNet-Affect, corresponding to the six basic emotion categories incorporated in a VSM. In the resulting model, terms were represented as vectors, with their components being the co-occurrence frequencies of words in document corpora. These frequencies had weights based on log entropy, in line with a TF-IDF weighting scheme. Kim et al.'s vector-based approach allowed the uniform representation of words, sentences, and WordNet-Affect synsets as vectors. They utilised the cosine angle between an input vector (representing an input sentence) and an emotional vector (representing an emotional synset) as a similarity measure to determine the emotion conveyed by a sentence. Kim et al. also set a predetermined threshold (t = 0.65) to affirm a strong emotional correlation between the two vectors.

Strapparava and Mihalcea (2008) developed five emotion analysis systems implementing different lexical and machine learning approaches. The goal was to classify newspaper headlines for Ekman's six emotions (anger, disgust, fear, joy, sadness, and surprise). The first system used the WordNet Affect lexicon to identify emotions in text. Next, they had the LSA Single Word system, which measured the similarity between the text and each emotion using LSA, with each emotion represented by a vector of a specific word, such as *anger*. The third system, LSA Emotion Synset, included synonyms from WordNet synsets for the emotion words. The fourth method, LSA All Emotion Words, incorporated all the words labelled with an emotion from all the WordNet Affect synsets. The fifth method was a Naive Bayes classifier trained on emotion-annotated blog data. The WordNet Affect system had high precision but low recall. The blog-based system performed best for joy and anger, likely due to more training data for these emotions, whereas the LSA models excelled in the other emotions.

Research in unsupervised machine learning suggests that conducting studies without relying on emotion lexicons is feasible. Kozareva et al. (2007) used Pointwise Mutual Information (PMI) to determine semantic relatedness, enhancing it with context-dependency rules. PMI was used to quantify the association between an entire phrase against a single emotion word. A high PMI value suggested a strong association between the two words or phrases, indicating that they often appear together more frequently than by random chance. Agrawal and An (2012) used PMI to evaluate Ekman's six basic emotions against a group of words representing each emotion and therefore considering the contextual setting of each emotion word. Their approach started with the extraction of NAVA words (Nouns, Adjectives, Verbs, and Adverbs) from each sentence. Following this, they extracted the syntactic dependencies between these words to incorporate contextual information into their model. They then employed semantic relatedness to calculate emotion vectors for these words. This is based on the premise that NAVA words frequently found together are likely to have semantic connections. They tested the model on three different non-emotion text corpora: Wikipedia data, which provides a broad range of content; Gutenberg corpus, consisting of over 36,000 e-books; and Wiki-Guten, a hybrid dataset created by merging the two corpora above. Additionally, they tested whether incorporating the syntactic dependency structure of sentences could offer contextual insights and thereby enhance emotion detection. The Wikipedia corpus achieved the highest F1-score, possibly due to its more structured data content. In contrast, the Gutenberg corpus, initially thought to contain more emotional text in its e-books, ranked last. The most extensive corpus, Wiki-Guten, did not yield the best results, indicating that performance varies across different methods and data sets. Interestingly, the F1-score indicated that the context-sensitive method was more appropriate for emotion classification tasks.

Calvo and Kim (2013) used the TF-IDF weighting approach to transform emotion words and sentences into vectors. The vectors are measures of the frequencies of word co-occurrences in the corpora. Calvo and Kim then applied a VSM model to represent words, sentences, and synsets as unified vectors. Furthermore, the VSM model data noise was reduced with Latent Semantic Analysis (LSA), Probabilistic LSA (PLSA), and Non-negative Matrix Factorization (NMF), categorizing the documents into categorical emotion labels. The study concludes that NMF-based categorical classification outperformed the other categorical classification methods. Zad and Finlayson (2020) replicated Kim et al. (2010) for detecting emotions in narrative text. This task involved an in-depth exploration of three distinct techniques for noise reduction or dimensionality reduction: Non-negative Matrix Factorization (NMF), Principal Component Analysis (PCA), and Latent Dirichlet Allocation (LDA). The results revealed that NMF was again the most effective method for detecting emotion categories. In the following we summarise the main unsupervised methods for detecting emotion categories, including their descriptions, applications, and limitations:

Latent Semantic Analysis (Deerwester et al., 1990)**:**

Description: Transforms terms or documents into a lower-dimensional space, simplifying complex data. Integrates with probabilistic theories for latent topic discovery.
Application: Effective in discovering hidden structures in texts and identifying associations between documents and topics.
Limitations: May miss specific emotional cues or sequences of expressions in the text.

Vector Space Model (Yates & Ribeiro-Neto, 1999):

Description: Represents text as vectors in a multidimensional space using the TF-IDF method and uses cosine similarity to determine the relationship between documents.
Application: Suitable for identifying broad patterns in texts and categorizing emotions based on word frequency and their co-occurrences.
Limitations: High-dimensional nature requiring dimensionality reduction techniques such as PCA or LSA; it may not capture subtle nuances of emotions.

Pointwise Mutual Information (Agrawal & An, 2012; Kozareva et al., 2007):

Description: Quantifies the association between phrases and emotion words, enhancing semantic relatedness with context-dependency rules.
Application: Useful for evaluating the contextual setting of each emotion word and identifying semantic connections.
Limitations: Requires large datasets; it may struggle with context disambiguation.

Non-negative Matrix Factorization: (Kim et al., 2010; Zad & Finlayson, 2020):

Description: Reduces data noise with a matrix factorization approach, categorizing documents into categorical emotion labels.

Application: Applicable for noise reduction and dimensionality reduction in emotion category detection.

Limitations: May not effectively represent the subtleties of emotions.

5.1.3 Unsupervised Methods for the Detection of Emotion Dimensions

Unsupervised machine learning methods are less used to detect text emotion dimensions. However, a few studies have ventured into this area. Two notable studies have applied NMF and PCA to explore the Valence, Arousal, and Dominance (VAD) emotion model in text data. The idea behind detecting emotion dimensions in texts is that by analysing and understanding the emotion space, we can narrow our focus to a few central dimensions rather than a wide range of individual emotions. These dimensions are crucial for research on the link between emotional states, actions, and behaviours. The use of unsupervised methods such as NMF and PCA highlights the potential of unsupervised learning techniques in emotion analysis despite the challenges posed by the high-dimensional nature of text data. This perspective is beneficial for comprehending human emotions and behaviours because it focuses on a limited set of core dimensions rather than an extensive array of categories.

Calvo and Kim (2013) used the database ANEW (Affective Norms for English Word; Bradley & Lang, 1999) to generate three-dimensional vectors of valence, arousal, and dominance. A sentence's VAD value was the same as the words' VAD values. Sentences were tagged with the nearest emotion category (anger, fear, disgust, joy, and sadness) in VAD space. This dimensional approach, emphasizing valence, arousal, and dominance, offered a psychologically motivated visualization of the emotion categories. Calvo and Kim evaluated three statistical dimensionality reduction techniques against the dimensional model: LSA, PLSA, and NMF. Calvo and Kim evaluated their dimensional (and categorical) methods using four datasets: SemEval, ISEAR, Fairy Tales, and USE. NMF excelled in both categorical and dimensional approaches. However, the results differed from one dataset to another, likely due to the texts from various domains and having different vocabularies.

Grgić et al. (2022) identified the VAD dimensions and uncovered novel dimensions from text, offering a more comprehensive explanation of basic emotions. Their primary objectives were to develop and validate a new method for identifying the space for mapping emotions. They gathered 16 datasets from various online public sources, focusing on social media and online reviews. These sources encompassed Kaggle,[1] a platform known for its numerous public datasets, Amazon review data from Julian McAuley,[2] and X-Twitter datasets collected following various hashtags. They then categorized the

---

[1] Kaggle.com/datasets.

[2] https://cseweb.ucsd.edu/~jmcauley/datasets.html#amazon_reviews.

datasets into three main groups: finance datasets, situational datasets featuring X-Twitter data related to various environments and situations, and online review datasets, which include data from reviews posted after purchasing products. Grgić et al. pre-processed the texts using two lexicons: the NRC Emotion Lexicon for assigning scores based on sentiments and basic emotions and the NRC Valence, Arousal, and Dominance Lexicon for valence, arousal, and dominance scores. After preprocessing, they applied PCA to extract orthogonal dimensions from each dataset, aiming for a 90% variance cutoff. The resulting dimensions, typically eight to ten per dataset, show consistency and independence across datasets. They then perform similarity clustering and global PCA analysis on these dimensions, analysing the correlation between dimensions across all datasets and forming a hierarchical structure of relationships. The PCA analysis involves combining matrices from each dataset to identify global orthogonal dimensions to ensure that the results are not biased by dataset size. PCA does not necessitate dividing data into training and test sets. Instead, in their PCA analyses Grgić et al. treated each dataset as a separate sample.

The findings revealed three consistent emotional dimensions across textual datasets. These included Valence, representing the positive–negative sentiment spectrum; Activation Arousal, combining arousal and dominance, challenging the traditional VAD model by suggesting dominance is not a standalone dimension; and Expectancy Tension, a new dimension related to anticipation and surprise, distinct from the core arousal dimension, focusing on emotional responses towards future events. Particularly, the Expectancy Tension dimension brings new insights into basic emotions tied to anticipation and surprise. Brief texts, such as social media posts and online reviews, can represent users' immediate emotional states as snapshots of human emotions. In the following, we have an overview of unsupervised methods for detecting emotion dimensions, including methods' descriptions, applications, and limitations.

Principal Component Analysis (Calvo & Kim, 2013):

Description: Extracts orthogonal dimensions from datasets, aiming for variance cutoff, and forms a hierarchical structure of relationships.
Applications: Suitable for identifying consistent emotional dimensions across textual datasets.
Limitations: Challenges posed by the high-dimensional nature of text data.

Global Principal Component Analysis (Grgić et al., 2022):

Description: Combines matrices from each dataset to identify global dimensions, ensuring results are not biased by dataset size.
Applications: Useful for mapping consistent emotional dimensions across diverse datasets.
Limitations: Complexity in handling and interpreting data from various domains.

5.2   Supervised Methods for Emotion Detection

Supervised machine learning models are engineered to use datasets annotated with emotion labels, lexicon, and corpora in which the emotion content has been labelled. This approach enables the supervised machine learning models to discern and learn patterns and features indicative of various emotion states.

Supervised methods start with collecting and preparing new datasets or using existing datasets annotated with emotion categories and/or dimensions. Feature extraction is a critical step in building a supervised model for emotion detection. Feature extraction involves transforming raw text into a format the algorithm can process, often using tokenization, stemming, and word embeddings. The resulting features represent the linguistic and semantic aspects of the text, providing a basis for the machine learning model to learn. We report the limitations of supervised and unsupervised methods for data quality, feature selection, model complexity, overfitting, ambiguous data handling, computational intensity, and scalability while detecting emotion categories and dimensions in the following:

o **Data Labelling**
   – *Supervised Models:* Requires labelled data (texts tagged with specific emotions)
   – *Unsupervised Models:* Works with unlabelled data (no explicit tagging of emotions)
o **Learning Process**
   – *Supervised Models:* Learning is guided by labels; models adjust based on prediction accuracy
   – *Unsupervised Models:* Models identify patterns or clusters without guidance from labels
o **Model Objective**
   – *Supervised Models:* Predict specific emotions for new texts based on learned patterns
   – *Unsupervised Models:* Explore data structure, identify clusters or patterns in texts
o **Applications**
   – *Supervised Models:* Suitable for categorizing texts into predefined emotional categories
   – *Unsupervised Models:* Useful for discovering new patterns and associations in text data
o **Complexity and Resources**
   – *Supervised Models:* Often requires substantial effort and resources for data labelling
   – *Unsupervised Models:* Less resource-intensive for data preparation but may require more analysis for interpretation.

As regards suitable machine learning models, common choices include decision trees, support vector machines (SVM), naive Bayes classifiers (NB), and neural networks, particularly those that are adept at handling sequential data, like recurrent neural networks (RNNs) and Long Short-Term Memory networks (LSTMs). The choice of machine learning model depends on the task's complexity, the dataset's size and nature, and the application's specific requirements. The machine learning model training involves feeding the model with a prepared dataset, allowing it to learn how different features correlate with specific emotion categories and/or dimensions. This phase is iterative; the model gradually improves its ability to predict emotions based on the feedback it receives from its performance on the training data. Once trained, the model's performance is evaluated using a separate test dataset. Metrics such as accuracy, precision, recall, and F1-score assess how well the model can identify emotions in new text samples. In the following, we provide a succinct description of each metric, allowing for easy comparison and understanding of their applications and limitations in the context of model evaluation. These metrics provide insight into the model's effectiveness and help fine-tune the parameters or select a different modelling approach (TP = True Positives: Correctly identified positives; TN = True Negatives: Correctly identified negatives; FP = False Positives: Incorrectly identified positives; FN = False Negatives: Incorrectly identified negatives).

Accuracy

Definition: Ratio of correctly predicted observations to the total observations.

$$Formula : \frac{(TP + TN)}{(TP + TN + FP + FN)}$$

Use-Case: Best for balanced datasets where each class is equally important.
Limitations: Can be misleading in imbalanced datasets.

Precision

Definition: Ratio of correctly predicted positive observations to the total predicted positives.

$$Formula : \frac{TP}{(TP + FP)}$$

Use-Case: When false positives are more problematic than false negatives. Useful in imbalanced datasets focusing on one specific class of emotions.
Limitations: Does not consider false negatives.

Recall

Definition: Ratio of correctly predicted positive observations to all observations in the actual class.

$$Formula : \frac{TP}{(TP + FN)}$$

Use-Case: When missing any positive case is critical. Important in scenarios where false negatives have serious implications.
Limitations: Ignores false positives.

F1-Score

Definition: Harmonic mean of precision and recall, providing a balance between the two.

$$Formula : \frac{2 * (Precision * Recall)}{(Precision + Recall)}$$

Use-Case: Useful in scenarios where both false positives and false negatives are equally important.
Limitations: May not be suitable if there is a need to prioritize precision or recall.

5.2.1 Supervised Methods for the Detection of Emotion Categories

Supervised machine learning classifiers operate with known inputs from annotated corpora and lexicons. Among the machine learning classifiers are Decision Tree (DT), k-Nearest Neighbour (K-NN), Support Vector Machine (SVM), Random Forest (RF), Naive Bayes (NB), Linear Regression (LR), and Multinomial Naive Bayes (MNB).

Decision Trees (DT) are a type of rule-based supervised classifier. They consist of a tree-like structure with nodes representing attributes or features, branches indicating decision rules on attributes, and leaf nodes denoting class labels or outputs (Hasan et al., 2019). Creating a decision tree involves three primary steps: selecting features, constructing the tree, and tree pruning. Kaur et al. (2012) showed that pruning, that is, removing less significant elements from the tree, can enhance DT performance by reducing overfitting.

The k-Nearest Neighbour (k-NN) algorithm uses labelled training data to predict the label of new data. For each new data point, k-NN looks at the k closest labelled data points (the k nearest neighbours) in the training set and assigns a label to the new point based on the majority label among these neighbours. Wu et al. (2008) noted that the k-NN algorithm commonly employs the Euclidean distance to calculate the proximity of an attribute to its neighbours.

Random Forest (RF) is another supervised classification algorithm developed from the decision tree method. It comprises a *forest* of decision trees combined to yield more precise and robust predictions (Fang & Zhan, 2015). This approach applies to both classification and regression problems. Segnini and Motchoffo (2019) discuss two main classification techniques used in RF: bagging and boosting. Bagging, or boot-strap aggregation, involves training N decision trees on different random subsets of data and averaging all predictions for the final output. Conversely, boosting strengthens the classifier by addressing errors from previous trees and forming a series of sequential predictors, effectively handling models that failed to predict accurately. Ensemble learning, which averages decisions from multiple predictors, tends to produce superior predictions compared to a single decision tree.

Support Vector Machine (SVM) is a supervised learning algorithm used predominantly for classification and regression tasks. As Duhan and Dhankhar (2016) outlined, SVM operates by creating hyperplanes in a high-dimensional space designed to classify data with the broadest possible margin, thereby enhancing the robustness of the classification. However, Zahid et al. (2021) noted a risk of overfitting, particularly when the number of features surpasses the number of samples.

The Naïve Bayes classifier (NB) is based on the Bayesian probability theorem, and it operates under the Naïve Bayes assumption, which posits that the features of an object are conditionally independent of each other, given the class label of the object. NB use the Bayes' rule to estimate the probability of features appearing in each class, subsequently identifying the most probable class. In the following, we will delve into various studies that have utilized the machine learning classifiers mentioned above. We will focus on exploring their application in emotion category detection in text.

Mohammad (2012) employed hashtags as labels for tweets, using SVMs to classify Ekman's six emotions. He demonstrated the effectiveness of hashtags over random classification.

Purver and Battersby (2012) utilised an SVM classifier on tweets and achieved 82% accuracy in identifying the emotion *happy*. Their approach used emoticons as training set labels and hashtags for the test set. They also tested the model ability to differentiate between various emotion classes, with accuracies ranging from 13 to 76%. Additionally, Purver and Battersby evaluated the effectiveness of using hashtags and emoticons as labels against a dataset of 1000 tweets labelled by human annotators. They found F-scores between 0.10 and 0.77 for different emotions. Their findings suggested that emoticons and hashtags could be an effective labelling strategy and a valid alternative to manual labelling.

Wang et al. (2012) constructed a dataset of approximately 2.5 million tweets tagged with emotion-related hashtags. They expanded the range of hashtags to encompass 131 words representing the seven basic emotions. The dataset quality was enhanced by selecting tweets that met specific criteria, such as being in English, containing minimal hashtags, and excluding URLs or quotations. Wang and colleagues experimented with various feature combinations on a subset of 250,000 tweets to determine the most effective set. This set included a mix of the LIWC lexicon, WordNet-Affect, and POS tagging. They then expanded the training dataset size and observed its impact on classification performance. The SVM classifier's outputs are binary vectors that indicate whether an instance possesses a specific label. These outputs were then the inputs for a Bayesian Network, which inferred final labels that achieved an F-measure ranging from 0.72 for joy to 0.13 for surprise. The variation in results was attributed to the unbalanced distribution and an overlap between emotions such as anger and sadness or joy and love in the training dataset.

Balabantaray et al. (2012) undertook a different approach to X-Twitter emotion classification, manually labelling approximately 8,000 tweets for six basic emotions defined in Ekman's model. They utilised an SVM multi-class classifier with 11 features, including Unigrams, Bigrams, pronouns, adjectives, and Word-net Affect lexicon. This approach achieved an accuracy of 73.24%.

Wikarsa and Thahir (2015) created a system to identify emotions in X-Twitter users using the NB method. They processed 105 tweets, converting them to lowercase, removing stop words, mentions, and URLs, and converting emoticons to text. Using NB, they categorised tweets into emotions such as happiness, sadness, and anger. The system's accuracy was 83%, but they suggested more training data and duplicate tweet removal for better results.

Jain et al. (2017) explored emotion extraction from multilingual texts across three domains. They employed a unique data collection method called rich site summary and used SVM and NB algorithms for classifying emotions in X-Twitter texts. Their findings indicated that NB achieved an accuracy of 71.4%.

Bučar et al. (2018) developed a Slovenian lexicon called JOB 1.0 and a labelled news corpus called SentiNews 1.0 includes 25,524 headwords, each rated on a sentiment scale from −5 to 5, similar to the AFINN lexicon. 10,427 documents were rated on a scale of 1 (negative) to 5 (very positive). The researchers found that the NB algorithm outperformed the SVM, achieving over 90% F1 score in binary classification and above 60% in sentiment classification.

Hasan et al. (2019) used NB, SVM, and Decision Trees for identifying emotions in text messages. Their study had two parts: the first involved collecting and automatically labelling a X-Twitter dataset using hashtags, followed by model training. The second part entailed developing a two-stage system called EmotexStream, which filters out emotionless tweets and then identifies emotions using the models trained in the first stage. This approach resulted in a 90% accuracy rate in emotion category classification.

Tiwari et al. (2020) applied SVM, Naïve Bayes, and Maximum Entropy using the n-gram feature extraction method on the Rotten Tomato dataset, which consisted of 1600 movie reviews for training and testing. They noted a decline in accuracy with larger n-grams. In the following list we summarise the supervised methods for detecting emotion categories reviewed so far.

Decision Tree (DT)—Hasan et al. (2019), Kaur et al. (2012):

Description: A tree-like structure with nodes representing features and leaf nodes denoting class labels. The method involves selecting features, constructing the tree, and tree pruning.
Findings/Applications: Kaur et al., 2012 demonstrated that pruning can enhance DT performance by reducing overfitting.

K Nearest Neighbour (KNN)—Wu et al. (2008):

Description: Utilizes labeled data to predict labels for new data, assigning labels based on the majority label among the k nearest neighbours.
Findings/Applications: Commonly employs Euclidean distance for calculating attribute proximity.

Support Vector Machine (SVM)—Mohammad (2012), Wang et al. (2012), Balabantaray et al. (2012), Jain et al. (2017), Tiwari et al. (2020):

Description: Commonly used for classification, it creates hyperplanes in a high-dimensional space for data classification.
Findings/Applications: Used in numerous studies for classifying emotions, with varying degrees of accuracy.

Random Forest (RF)—Fang and Zhan (2015), Segnini and Motchoffo (2019):

Description: An ensemble of decision trees combined to yield more accurate predictions.
Findings/Applications: Produces superior predictions compared to a single decision tree.

Naïve Bayes (NB)—Purver and Battersby (2012), Wikarsa and Thahir (2015), Bučar et al. (2018):

Description: Based on the Bayesian probability theorem, it assumes conditional independence of features given an emotion category.

Findings/Applications: Employed in different studies for emotion classification in tweets with varying accuracy rates.

5.2.2 Supervised Methods for the Detection of Emotion Dimensions

Supervised methods can offer a deep understanding of how language and context influence emotional expressions, leading to more accurate emotion detection than unsupervised or lexicon-based methods. However, their use in detecting emotion dimensions has been somewhat limited. Developing supervised models for this purpose requires large datasets accurately labelled for emotion dimensions. Although there are lexicons and corpora annotated for emotion dimensions in the literature, they are much less common than those annotated for basic emotion categories. Another challenge with categorising emotion dimensions is their intrinsic complexity; they involve more than just categorising emotions. These models need to detect subtle variations in valence, intensity, dominance, and context. Despite these challenges, some studies have undertaken the task of modelling emotion dimensions using supervised learning techniques.

Hasan et al. (2013) devised a method to analyse tweets for emotional content using a feature set based on the Circumplex emotion model. They employed four different classifiers—Naive Bayes, SVM, Decision Tree, and KNN—and achieved an accuracy of approximately 90% in emotion classification. Building on this work, Hasan et al. (2019) developed an advanced system to detect emotions in streams of tweets. The system employed a two-tier classification approach and was trained using the lexicons ANEW, LIWC, and AFINN. They first identified the tweets for valence and intensity and then further categorised them into specific emotional categories. The emotional categories captured by the final system included the valence (positive–negative) dimension, the emotions of anger and sadness, the state of anxiety, and the negations.

Mashal and Asnani (2017) used NB and SVM to examine emotion intensity on X-Twitter data. The data were collected automatically using emotion-related hashtags, covering five emotions: happiness, sadness, anger, fear, and surprise. The results indicate that the performance of intensity detection by the classifiers was consistent, even with larger datasets.

Soumya and Pramod (2020) analysed 3,184 Malayalam tweets, categorising them into positive and negative polarity, using different feature vectors, including BOW and SentiWordNet. They implemented RF and NB, observing that RF, particularly when combined with SentiWordNet and negation words, was the most effective, achieving an accuracy of 95.6%. The following list summarises the supervised methods detecting emotion dimensions and their findings reviewed so far.

Naïve Bayes (NB)/Support Vector Machine (SVM)—Hasan et al. (2013), Mashal and Asnani (2017):

Description: Employed for analysing emotional content in tweets, focusing on emotion intensity and valence.
Findings/Applications: Achieved 90% accuracy in emotion classification, with consistent performance in intensity detection.

Random Forest (RF)—Soumya and Pramod (2020):

Description: Used for categorizing tweets into positive and negative polarity, with a focus on dimensions like sentiment.
Findings/Applications: Highly effective, especially when combined with SentiWordNet and negation words, achieving an accuracy of 95.6%.

Decision Tree (DT)/K Nearest Neighbour (KNN)—Hasan et al. (2013), Hasan et al. (2019):

Description: Part of a multi-classifier system for detecting emotions in tweets, with a focus on various emotional dimensions like valence.
Findings/Applications: In two-tier classification approaches, it reaches high accuracy.

5.3 Limitations of Unsupervised and Supervised Machine Learning Methods for Emotion Detection

In this section, we contrast the approaches of supervised and unsupervised machine learning and highlight the main limitations of these methods for emotion detection. While unsupervised learning offers the advantage of working without annotated datasets, the success of unsupervised learning models heavily depends on the quality and structure of the input data. The unsupervised model may struggle to find relevant patterns if the data is noisy. The selection of features also plays a critical role in unsupervised learning: choosing the wrong set of features can lead the model to focus on irrelevant data, resulting in misleading patterns. Unlike supervised learning, where the model's performance can be directly measured against a labelled dataset, the absence of predefined labels in unsupervised learning makes it more challenging to assess the validity of the patterns identified by the model.

The limitations of supervised and unsupervised methods for data quality, feature selection, model complexity, overfitting, ambiguous data handling, computational intensity, and scalability while detecting emotion categories and dimensions are reported in the following:

o **Data Requirement**
  - *Unsupervised Methods:* Does not require labelled data but needs large datasets for effective pattern recognition
  - *Supervised Methods:* Requires large datasets labelled with emotions categories and/ or dimensions
o **Data Preprocessing**
  - *Unsupervised Methods:* Heavily reliant on quality and structure of input data for effective pattern discovery
  - *Supervised Methods:* Requires careful preprocessing to ensure quality input data for model training
o **Feature Engineering**
  - *Unsupervised Methods:* Less dependent on feature engineering but may require preprocessing to structure data for pattern discovery
  - *Supervised Methods:* Needs careful feature selection and extraction
o **Model Complexity**
  - *Unsupervised Methods:* Often relies on simpler models, which may not capture the complexity of human emotions effectively
  - *Supervised Methods:* Models can become complex and computationally intensive, especially for nuanced emotion detection
o **Ambiguity**
  - *Unsupervised Methods:* Difficulty in interpreting complex emotional nuances and contextual subtleties without labelled guidance
  - *Supervised Methods:* Struggles with the subjective and nuanced nature of emotions
o **Context Disambiguation**
  - *Unsupervised Methods:* Can capture broader patterns but may miss specific cues or the sequence of expressions in text
  - *Supervised Methods:* Better at capturing specific emotion expressions but may miss broader context in text
o **Generalization**
  - *Unsupervised Methods:* May find general patterns but can struggle to apply these effectively to specific or nuanced emotional expressions
  - *Supervised Methods:* Risk of overfitting to training data, leading to poor performance on test dataset
o **Scalability**
  - *Unsupervised Methods:* More scalable in terms of data requirements but may face limitations in model complexity and interpretability
  - *Supervised Methods:* Scalable with increased data but requires more labelled data and computational resources
o **Emotion Categories**
  - *Unsupervised Methods:* May struggle to distinctly categorize emotions without clear labels, leading to ambiguous or overlapping emotion clusters

- *Supervised Methods:* Effective in identifying distinct emotion categories but requires accurately labelled examples for each category
o **Emotion Dimensions**
  - *Unsupervised Methods:* Finds patterns in dimensions without explicit labelling but may inaccurately represent the subtleties of the dimensions
  - *Supervised Methods:* Can be tailored to detect dimensions (valence, arousal) but requires labelled data representing these dimensions

Unsupervised models are not immune to overfitting, mainly if they are very complex or the data lacks diversity, leading them to capture noise instead of meaningful information. Interpreting the results can be challenging since there are no predefined labels to guide the understanding of the patterns or groupings identified by the model, leading to difficulties in drawing meaningful insights and conclusions. Additionally, regarding computational demands, some unsupervised algorithms, especially those handling large datasets, can be resource intensive. Consequently, scaling unsupervised models to handle large datasets presents challenges due to computational constraints and the difficulty in maintaining pattern quality.

As regards supervised machine learning methods, we can also find some limitations, especially when applied to emotion detection. The primary challenge of supervised methods is their heavy reliance on annotated datasets, which are time-consuming to prepare and can be somewhat ambiguous due to the subjective nature of emotions. Additionally, supervised methods often require extensive feature engineering, a process where we need to identify and extract relevant features from text. Moreover, supervised models tend to overfit. They also struggle with generalization, scalability, feature engineering, model complexity, and depend on preprocessing.

The data preprocessing dependence highlights how sensitive a model is to the data preparation. Such preparation includes cleaning the data, normalizing the features, handling the missing values, and other transformations. The performance of some models is highly contingent on the quality and method of data preprocessing, making this step critical. Feature engineering is the process of selecting and transforming raw data into meaningful model inputs (or features). In this step, we identify the most critical aspects of the data and how to structure these aspects for the model. Effective feature engineering can significantly improve the model's accuracy and efficiency. The model complexity refers to the structural intricacy of a machine learning model, often characterized by the number of its parameters. Complex models can capture intricate patterns but can be more prone to overfitting and require substantial computational resources. Conversely, simpler models are more computationally efficient and less likely to overfit but may not capture complex patterns effectively. The generalization is the ability of a model to perform effectively on new data. This trait indicates the model's capability to apply learned patterns to different datasets. A model that generalizes well can accurately predict or make decisions across various scenarios, avoiding the pitfalls of overfitting where performance is only strong on familiar data. Lastly, scalability refers to the model's capacity to handle an increasing volume of data. A scalable model should maintain its performance as the data volume expands.

In this Chapter we have discussed the pivotal role of machine learning in advancing emotion detection in NLP. We introduced various algorithms and showed their impact on improving the precision of emotion classification. However, while both supervised and unsupervised machine learning methods have advanced the field significantly, they come with certain limitations. Supervised learning relies heavily on annotated datasets, which can be time-consuming and expensive to create. Additionally, these models can only work with generalisation if the training data adequately represents the diversity of real-world language use. Unsupervised learning, while not dependent on labelled data, often falls short in accurately capturing the specific context of emotion expressions, as it lacks the guidance provided by annotated examples.

Neural networks and deep learning methods helped overcome these limitations. Due to their ability to process large amounts of data, neural networks and deep learning models excel in understanding context and handling language subtleties. They can learn and generalise across different types of textual data, reducing the reliance on extensive labelled datasets. These methods also do not need extensive manual intervention, making them more adaptable and scalable. In Chap. 6, we will look at neural networks, deep learning and transformer models in more detail, explaining their basic principles and how they are used for emotion detection in textual data.

References

Agrawal, A., An, A., & Papagelis, M. (2018). Learning emotion-enriched word representations. In *Proceedings of the 27th International Conference on Computational Linguistics* (pp. 950–961).

Agrawal, A., & An, A. (2012). Unsupervised emotion detection from text using semantic and syntactic relations. *Proceedings of the International Joint Conferences on Web Intelligence and Intelligent Agent Technology, 1*, 346–353.

Aman, S., & Szpakowicz, S. (2007). Identifying expressions of emotion in text. In V. Matoušek, & P. Mautner (Eds.) *Text, Speech and Dialogue*. Lecture Notes in Computer Science, Vol. 4629. Springer.

Balabantaray, C. R., Mudasir, M., & Nibha, S. (2012). Multi-class Twitter emotion classification: A new approach. *International Journal of Applied Information Systems, 4*, 48–53.

Bradley, M.M., & Lang, P.J. (1999). *Affective norms for English words (ANEW): Instruction manual and affective ratings*. https://api.semanticscholar.org/CorpusID:145474983

Bučar, J., Žnidaršič, M., & Povh, J. (2018). Annotated news corpora and a lexicon for sentiment analysis in Slovene. *Language Resources & Evaluation, 52*, 895–919.

Calvo, R. A., & Kim, S. M. (2013). Emotions in text: Dimensional and categorical approaches. *Computational Intelligence, 29*(3), 527–543.

Charles, W. G. (2000). Contextual correlates of meaning. *Applied Psycholinguistics, 21*(4), 505–524.

Deerwester, S. C., Dumais, S. T., Landauer, T. K., et al. (1990). Indexing by latent semantic analysis. *Journal of the American Society for Information Science, 41*, 391–407.

Duhan, B., & Dhankhar, N. (2019). Hybrid approach of SVM and feature selection based optimization algorithm for big data security. In *Proceedings of ICETIT*.

Fang, X., & Zhan, J. Z. (2015). Sentiment analysis using product review data. *Journal of Big Data, 2*, 1–14.

Firth, J. R. (1957). A synopsis of linguistic theory, 1930–1955. In J.R. Firth (Ed.), *Studies in Linguistic Analysis*. Blackwell.

Giatsoglou, M., Vozalis, M., Diamantaras, K., Vakali, A., Sarigiannidis, G., & Chatzisavvas, K. (2017). Sentiment analysis leveraging emotions and word embeddings. *Expert Systems with Applications, 69*, 214–224.

Grgić, D., Podobnik, V., & Carvalho, A. (2022). Deriving and validating emotional dimensions from textual data. *Expert Systems with Applications, 198*.

Hasan, M., Rundensteiner, E., & Agu, E. (2013). EMOTEX: Detecting emotions in Twitter messages. In *ASE BIGDATA/SOCIALCOM/CYBERSECURITY Conference*.

Hasan, M., Rundensteiner, E., & Agu, E. (2019). Automatic emotion detection in text streams by analyzing Twitter data. *International Journal of Data Science and Analytics, 7*, 35–51.

Jain, V. K., Kumar, S., & Fernandes, S. L. (2017). Extraction of emotions from multilingual text using intelligent text processing and computational linguistics. *Journal of Computational Science, 21*, 316–326.

Kaur, S., Rhati, M., & Mamta, A. (2012). Decision tree: Data mining techniques. *International Journal of Latest Trends in Engineering and Technology, 1*(3), 150–155.

Khosla, S., Chhaya, N., & Chawla, K. (2018). Aff2Vec: Affect–enriched distributional word representations. In *Proceedings of the 27th International Conference on Computational Linguistics* (pp. 2204–2218).

Kim, S. M., Valitutti, A., & Calvo, R. A. (2010). Evaluation of unsupervised emotion models to textual affect recognition. In *Proceedings of the NAACL HLT 2010 Workshop on Computational Approaches to Analysis and Generation of Emotion in Text* (pp. 62–70).

Kort, B., Reilly, E., & Picard, R. W. (2001). An affective model of interplay between emotions and learning: reengineering educational pedagogy-building a learning companion. In *Proceedings IEEE International Conference on Advanced Learning Technologies* (pp. 43–46).

Kozareva, Z., Navarro, B., Vázquez, S., & Montoyo, A. (2007). UA-ZBSA: A headline emotion classification through web information. In *Proceedings of the Fourth International Workshop on Semantic Evaluations (SemEval-2007)* (pp. 334–337).

Kusal, S., Patil, S., Kotecha, K., Aluvalu, R., & Varadarajan, V. (2021). AI based emotion detection for textual big data: Techniques and contribution. *Big Data Cognitive Computing, 5*, 43.

Mashal, S. X., & Asnani, K. (2017). Emotion intensity detection for social media data. In *International Conference on Computing Methodologies and Communication* (pp. 155–158).

Mikolov, T., Chen, K., Corrado, G., & Dean, J. (2013). *Efficient estimation of word representations in vector space*. ArXiv

Mohammad, S. M. (2012). #Emotional tweets. *Proceedings of the First Joint Conference on Lexical and Computational Semantics, 1*, 246–255.

Mohammad, S. M., & Turney, P. D. (2013). Crowdsourcing a word-emotion association lexicon. *Computational Intelligence, 29*, 436–465.

Pennington, J., Socher, R. & Manning, C. (2014). GloVe: Global Vectors for word representation. In *Proceedings of the 2014 Conference on Empirical Methods in Natural Language Processing* (pp. 1532–1543).

Plutchik, R. (1980). A general psychoevolutionary theory of emotion. In R. Plutchik & H. Kellerman (Eds.), *Theories of Emotion* (pp. 3–33). Academic press.

Purver, M., & Battersby, S. (2012). Experimenting with distant supervision for emotion classification. In *Proceedings of the 13th Conference of the European Chapter of the Association for Computational Linguistics* (pp. 482–491).

Segnini, A., & Motchoffo, J.J.T. (2019). *Random forests and text mining.* https://www.academia.edu/11059601/Random_Forest_and_Text_Mining

Soumya, S., & Pramod, K. V. (2020). Sentiment analysis of Malayalam tweets using machine learning techniques. *ICT Express, 6*(4), 300–305.

Tang, D., Wei, F., Qin, B., Yang, N., Liu, T., & Zhou, M. (2016). Sentiment embeddings with applications to sentiment analysis. *IEEE Transactions on Knowledge and Data Engineering, 28*(2), 496–509.

Tiwari, P., Mishra, B. K., Kumar, S., & Kumar, V. (2020). Implementation of n-gram methodology for rotten tomatoes review dataset sentiment analysis. In *Cognitive Analytics: Concepts, Methodologies, Tools, and Applications* (pp. 689–701).

Valitutti, A., Strapparava, C., & Stock, O. (2004). Developing affective lexical resources. *Psychology Journal, 2*(1), 61–83.

Wang, W., Chen, L., Thirunarayan, K., & Sheth, A. P. (2012). Harnessing twitter "big data" for automatic emotion identification. In *Privacy, Security, Risk and Trust International Conference* (pp. 587–592).

Wang, S., Maoliniyazi, A., Wu, X., & Meng, X. (2020). Emo2Vec: learning emotional embeddings via multi-emotion category. *ACM Transaction of Internet Technology, 20*(2), 17.

Wikarsa, L., & Thahir, S.N. (2015). A text mining application of emotion classifications of Twitter's users using Naïve Bayes method. In *1st International Conference on Wireless and Telematics* (pp. 1–6).

Wu, X., Kumar, V., Ross Quinlan, J., et al. (2008). Top 10 algorithms in data mining. *Knowledge Informatic Systems, 14*, 1–37.

Yates, R., & Ribeiro-Neto, B. (1999). *Modern information retrieval.* ACM Press.

Zad, S., & Finlayson, M. (2020). Systematic evaluation of a framework for unsupervised emotion recognition for narrative text. In *Proceedings of the First Joint Workshop on Narrative Understanding, Storylines, and Events* (pp. 26–37).

Zahid, R., Idrees, M. O., Mujtaba, H., & Beg, M. O. (2021). Roman Urdu reviews dataset for aspect-based opinion mining. In *Proceedings of the 35th IEEE/ACM International Conference on Automated Software Engineering, Association for Computing Machinery* (pp. 138–143).

Deep Learning and Transformers for Emotion Detection

6

6.1 An Overview on Neural Networks and Deep Learning Models

In the previous Chapter, we have seen that supervised and unsupervised machine learning methods have specific limitations. Supervised learning needs annotated datasets, which are costly and time-consuming to create. It also relies on diverse training data to generalizel. Unsupervised learning, which doesn't require labelled data, often struggles with accurately capturing the context of emotional expressions. Additionally, while these methods can be relatively quick to train on small datasets, they can become slower with larger and more complex data.

Neural networks, and particularly deep learning, try to overcome these issues in different ways. Neural networks are inspired by the human brain and excel at recognizing patterns and making decisions. Although training these models can be time-consuming and computationally intensive, they are capable of processing large amounts of data, understanding context, and handling language subtleties better than traditional methods. Once trained, they can make predictions very quickly. In this Chapter, we will explore the basics of deep learning and neural networks and see how they are specifically applied to improve emotion detection systems. We will then focus on transformers and GPT models, and their capacity on recognising emotions and even mimic empathy.

Neural networks consist of layers of interconnected *neurons* called nodes. Each neuron receives input, processes it, and passes on its output to the neurons of the next layer. The perceptron is the simplest form of a neural network, consisting of a single layer. On the other hand, deep learning involves the use of multiple, or *deep*, layers, enabling the handling of complex tasks and datasets. The layers are not directly exposed to the input or output but perform most computations. The number of layers and neurons within them can vary, greatly influencing the network's complexity and ability to handle intricate tasks.

© The Author(s), under exclusive license to Springer Nature Switzerland AG 2025   75
F. Cavicchio, *Emotion Detection in Natural Language Processing*, Synthesis Lectures
on Human Language Technologies, https://doi.org/10.1007/978-3-031-72047-5_6

Weights and biases are the parameters that the network adjusts during its learning process. Weights are tied to the connections between neurons across different layers, dictating the strength and influence of these connections. In contrast, the biases are added to the sum of the neuron's weighted input. The fundamental components of a neural network are:

Neurons (Nodes)
- Basic units of a neural network, mimicking neurons in the brain.
- Each neuron receives the input, processes it, and sends the output to the next layer.

Layers
- Input Layer: Receives data.
- Hidden Layers: Process data via weighted connections.
- Output Layer: Presents result/output.

Weights and Biases
- Connection between neurons has an associated weight and bias.
- Weights and bias are adjusted during the learning process to improve predictions.

Figure 6.1 shows a schematic organization of the fundamental components of a neural network.

A neural network includes several computational processes, starting with forward propagation. In this phase, the data is fed into the input layer and moves through the network.

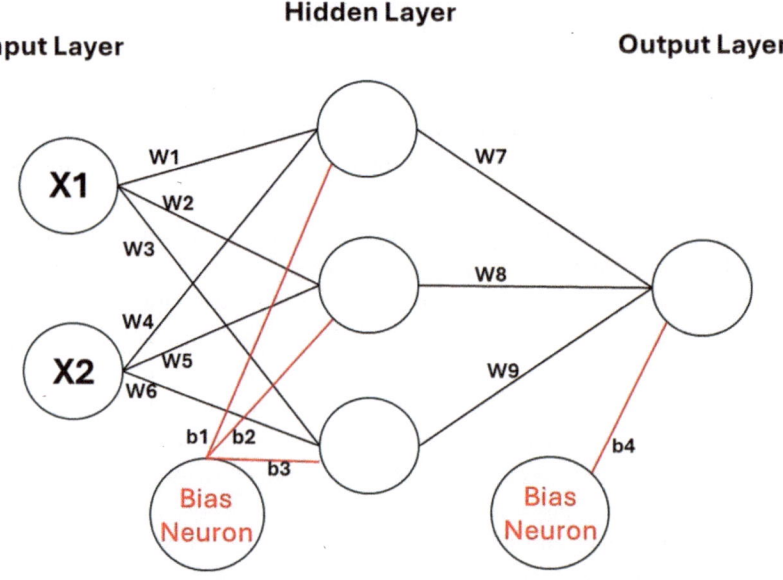

Fig. 6.1 Schematic representation of the neurons, layers, weights, and biases in a shallow neural network

Each neuron in the network multiplies the input it receives by a weight, adds a bias, and then passes this value through an activation function. The result of this computation is then forwarded to the next layer of neurons. This process continues until it reaches the output layer, providing the network's final output.

The activation function introduces non-linearity into the network. Non-linearity is crucial because it allows the network to learn and model complex patterns in the data. Two popular activation functions are the Sigmoid and the ReLU (Rectified Linear Unit). The Sigmoid function maps any real-valued input to a value between 0 and 1, making it particularly useful for models that predict probabilities, such as in binary classification problems. The ReLU function, on the other hand, outputs the input directly if it has a positive value and assigns zero to negative inputs. ReLU, which is partially linear but effectively introduces non-linearity, is usually favoured for its computational efficiency and ability to mitigate the attenuation of the gradient in deep networks, i.e., the vanishing gradient problem.

A key component of neural networks is the backpropagation algorithm, which adjusts the weights of the connections between the nodes in the network based on the output error compared to the expected result. This learning process is iterative, involving multiple cycles of forward and backpropagation. During each cycle, the network processes the input data, calculates the output, measures the error, and then adjusts its weights and biases to minimize this error in subsequent iterations. The backpropagation algorithm typically uses methods like gradient descent to guide these adjustments.

The learning process of a neural network also involves training and testing. During training, the network is exposed to a dataset and learns by iteratively adjusting its weights and biases through multiple cycles of forward and backpropagation. The network's learning ability is then evaluated by testing its performance on a new test dataset, assessing its ability to generalize and make accurate predictions or decisions. In the following, we report and briefly describe the fundamental processes of a neural network:

Forward Propagation
- Data passes through layers from input to output, with each input multiplied by its weight and summed with a bias, then passed through an activation function.

Activation Functions
- Functions such as Sigmoid or ReLU introduce non-linearity, allowing the network to learn complex tasks.

Learning Process
- Involves feeding large datasets to the network, allowing it to adjust weights and biases iteratively to reduce the error.

Backpropagation
- The output error is propagated back through the network, adjusting weights and biases to minimize the error.

Training and Testing
- The networks are trained on a subset of data, then tested on a separate dataset to evaluate performance.

Deep learning, an advanced branch of machine learning, excels in word embeddings. Word embeddings convert words from the language into multi-dimensional vectors. The vectors capture a wealth of semantic and syntactic information about the words, reflecting their meanings and contextual relationships. For example, words with similar meanings are located close to each other in the vector space, allowing the model to understand synonyms and theme similarities. Another strength of deep learning is its ability to learn hierarchical representations, meaning it can recognize patterns at multiple levels of abstraction, from simple features to complex linguistic structures. Integrating word embeddings into deep learning frameworks provides a more nuanced and contextually aware machine understanding of language.

Figure 6.2 shows an example of a neural network during learning.

Convolutional Neural Networks (CNNs), Recurrent Neural Networks (RNNs), and, more recently, Transformer models are prominent deep learning approaches for processing and classifying text data. CNNs utilize a mathematical operation called convolution, similar to multiplying two functions, employing convolution filters. The filters, often square

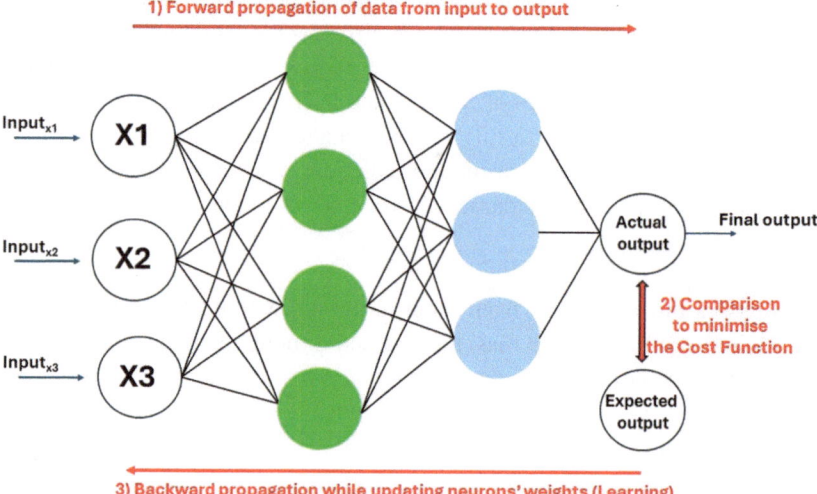

Fig. 6.2 Illustration of the learning process in a neural network. During each epoch, labelled data is fed from the input layer to the output layer (1), where the network's output is compared to the expected output (2). The difference, known as the cost function, is minimized through backpropagation, which adjusts the weights of the neurons (3). This process is repeated over multiple epochs, gradually reducing the loss function, and improving the network's accuracy

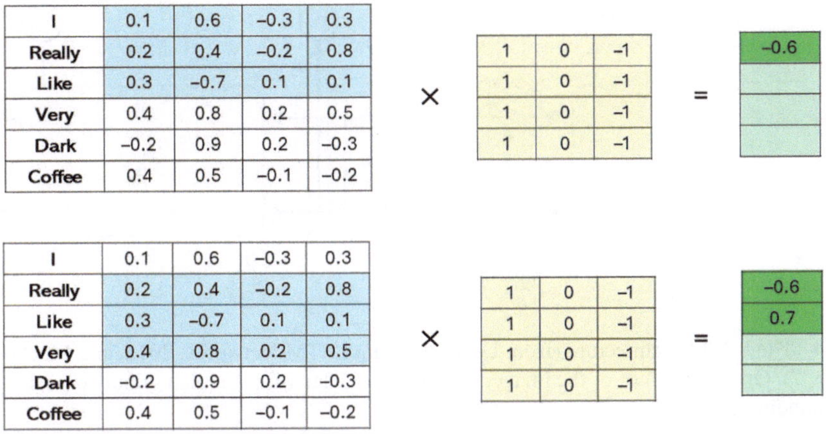

I	0.1	0.6	−0.3	0.3
Really	0.2	0.4	−0.2	0.8
Like	0.3	−0.7	0.1	0.1
Very	0.4	0.8	0.2	0.5
Dark	−0.2	0.9	0.2	−0.3
Coffee	0.4	0.5	−0.1	−0.2

×

1	0	−1
1	0	−1
1	0	−1
1	0	−1

=

−0.6

I	0.1	0.6	−0.3	0.3
Really	0.2	0.4	−0.2	0.8
Like	0.3	−0.7	0.1	0.1
Very	0.4	0.8	0.2	0.5
Dark	−0.2	0.9	0.2	−0.3
Coffee	0.4	0.5	−0.1	−0.2

×

1	0	−1
1	0	−1
1	0	−1
1	0	−1

=

−0.6
0.7

Fig. 6.3 A schematic illustration of the convolutional procedure converting the text into a vector format (Machová et al., 2023, p. 7)

matrices, are integral to convolutional layers. In text data processing, CNNs use a convolutional filter moving in a single direction and covering the entire length of a word vector (Machová et al., 2023). For neural network processing, the text must be converted into numerical form, specifically into vector representations (see Fig. 6.3).

RNNs is widely used discern relationships between words. These models treat text as a sequence, maintaining word order and context. Regarding, Long Short-Term Memory (LSTM) network is widely used in various classification tasks (Ghosh et al., 2020). LSTMs address the vanishing gradient problem by effectively retaining information over longer sequences. The integration of repeating LSTM blocks arranged in a chain helps to process vectorized word information throughout LSTM, the neuron structure, which includes the sigmoidal activation function (Wang et al., 2016). Moreover, adding attention mechanisms in LSTM models effectively extracts relevant information from long sentences (Li et al., 2020; see Fig. 6.4 for a schematic representation of the LSTM architecture).

Gated Recurrent Unit (GRU), introduced by Chung et al. (2014) and Farzi and Bolandi (2016), is an efficient alternative within the RNNs for gating. GRUs streamline the architecture of LSTM models, incorporating two gates and omitting the internal memory component, thereby simplifying the structure. Cho et al. (2014) emphasized that GRUs address the RNNs' vanishing gradient issue, modifying the LSTM with an update and reset gate strategy. The gates determine the proportion of past information to retain or discard in future sequences. The Bi-GRU extends this architecture, allowing bidirectional processing.

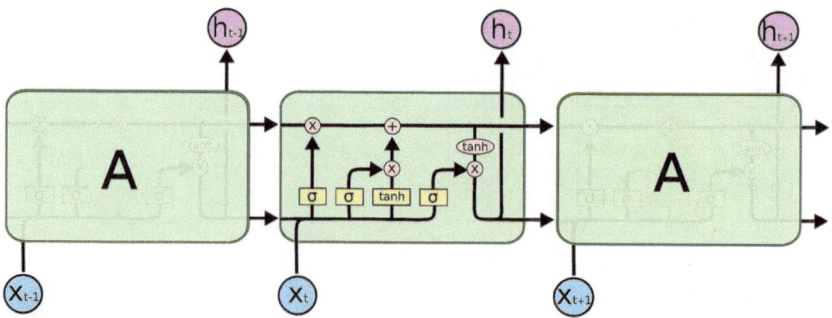

Fig. 6.4 The LSTM architecture (from Understanding LSTM Networks, 2015, and Machová et al., 2023, p. 7). Original creator of the image, Christopher Colah (https://colah.github.io/posts/2015-08-Understanding-LSTMs/)

6.1.1 Deep Learning for the Detection of Emotion Categories

We have seen in previous Chapters that the keyword-based approach is a prevalent method for detecting emotions in text. In deep learning, we can leverage the neural network layers to discern relevant keywords from labelled and unlabelled emotion datasets. The process usually starts with tokenisation, removing stop words, and lemmatising the emotion dataset. After the dataset treatment, embeddings are created by assigning numerical values to the tokens. We then apply classification algorithms, inputting the numeric vectors into the deep neural network. Within the network, layers are aligned with corresponding emotion labels. Thus, the network learns to recognise patterns in the data, which it then applies to predict the associated emotion labels. For example, Shrivastava et al. (2019) presented a novel annotated dataset derived from transcripts of a TV show annotated using Ekman's six basic emotions. The study employed a sequential CNN, which sequentially processes training data for emotion detection. This approach enables the network to leverage the characteristics of preceding sentences for emotion detection in subsequent ones. The initial layer of the network uses pre-trained word embeddings to transform words into vectors, ensuring that words with similar meanings have similar vector representations. The network's ability to grasp semantic and contextual nuances is further enhanced by feeding context features into an attention model. The attention model assesses the relevance of the context in the analysis. Integrating the attention mechanism with CNN allows the model to selectively focus on the most informative features within the input sequences. The attention mechanism in the proposed model mainly aids the CNN in concentrating on words or features that significantly impact the classification.

The performance of the CNN model is evaluated against LSTM networks and Random Forest classifiers. The study focused on training accuracy for both 6 Ekman's emotions, plus the neutral state, and 3-emotion classifications, revealing that the CNN model surpassed both LSTM and RF classifiers with accuracies of 80.41% for fine-grained emotions such as *anger*, *fear* and *happiness* and 83.32% for the coarse-grained emotions (positive or negative emotions and the neutral state). The CNN model demonstrated a 77.54% accuracy for fine-grained emotions and 80.99% for coarse-grained emotions. Notably, the precision for emotions such as *anger* and *fear* improved after addressing class imbalance, while *happiness* showed better recall and F1-score. Despite the predominance of the neutral category, the negative class exhibited high precision and F1-score, indicating reduced impact from the neutral class, which accounts for the 38.1% across the emotions. Although the model accurately predicted most true positive classes, there was some confusion between emotions such as *anger* and *disgust* on one side and *happiness* and *surprise* on the other, possibly due to semantic similarities in the words used for these emotions. Shrivastava et al.'s study also highlighted the efficacy of the attention mechanism in selecting keywords with the phrase *go away* being given significant weight. Although the text was classified under *anger*, a slight overlap with the *surprise* category was observed, underscoring the intricate nature of emotion classification in text.

In their study, Liu et al. (2021) introduced their Semantic and Emoticon Based Emotion Recognizer (SEER), an innovative model for identifying emotions in text. This model combines a Bi-GRU network with an attention mechanism to determine emotion vectors precisely based on the aspects of input words. Additionally, the model incorporates an analysis of emoticon distribution from the dataset, utilising this information to enhance the emotion vectors. SEER is structured to classify emotions in short texts into categories such as *joy*, *sadness*, *anger*, *fear*, *disgust*, or *surprise*. The SEER model comprises several components:

- A word embedding layer, a Bi-GRU neural network for extracting semantic features.
- An attention mechanism to assign varying weights to each word.
- An emoticon embedding to interpret emotions in emoticons.
- A connection layer that mixed semantic and emoticon features for emotion detection.

Liu et al. also compared different methods, including lexicon-based approaches, machine learning, and deep learning techniques. Their findings revealed that machine-learning methods outperform lexicon-based ones, and deep-learning approaches show the highest efficacy. Specifically, Bi-GRU and Bidirectional LSTM outperformed their unidirectional counterparts. Including an attention mechanism and emoticon emotion analysis in the SEER model further enhanced the model performance.

Basile et al. (2019) explored the detection of emotions in English textual exchanges between humans and chatbots. The language used in the annotated dataset exhibited characteristics typical of micro-blogging platforms, including widespread use of contractions (such as *I'm gonna bother*), elongations (*a vacation tooooooo!*), non-standard punctuation (*gonna explain you later..!*), and intentional or unintentional misspellings (*U r*). They randomly chose 2754 samples from the official training set, adhering to the class distribution specified for the official development and test sets. This selection led to two datasets consisting of 110 sentences each for the *sad*, *angry*, and *happy* categories and 2,424 instances for the *others* class, respectively. Basile et al. designated these datasets as int-Dev and intTest sets, respectively, while the remaining portion of the training dataset was referred to as intTrain. Basile et al. then developed four distinct neural network models for their SemEval 2019 application. In their first approach, Basile et al. handle the three turns of the human-chatbot conversation by allocating a separate input branch for each and explicitly marking the position of each sequence within the dialogue. These branches used word embeddings and fed into a two-layer bidirectional LSTM with an attention mechanism. This synthesis allows for independent processing of the three conversational turns while focusing on the most pertinent parts of each. The gathered information from these branches is then merged through simple concatenation followed by several connected dense layers. This model aimed to facilitate classifying emotion versus *other* categories. The second model adapted to the imbalanced nature of the dataset, which was predominantly of the *others* class. Basile et al. employed a multi-task learning strategy that included an auxiliary output that merged the labels *angry*, *happy*, and *sad* into a single emotional category. The third model used transfer learning with a straightforward feed-forward network, enhancing it with a fine-tuned Universal Sentence Encoder (Cer et al., 2018). For the input, they chose the first and last turns of the conversation. This decision was based on the observation that incorporating the second turn (the bot's turn) into the model resulted in diminished performance. Finally, Basile et al. fine-tuned a BERT-base model, treating the problem as a sentence-pairs issue. In their approach, they utilised the first and third conversational turns as a pair, intentionally omitting the bot's turn. The model was further enhanced by integrating a lexical normalisation system. The four systems' performance was assessed using precision and recall for each emotion category, along with the F1-score. Basile et al. observed that the best neural system was the third, and it achieved an F1-score of 0.73 on the intTest set and 0.72 on the official SemEval test set. They noted a significantly higher recall for the *happy* and *sad* classes on the intTest set than the official development and test sets. Conversely, the two top-performing neural systems (the third and the second) showed lower precision for these classes on the intTest dataset than on the official datasets.

In their submission to SemEval 2018, Ezen-Can and Can (2018) introduced a method that employed a series of binary classifiers, each dedicated to one of the 11 emotion categories of Mohammad et al.'s (2018) annotated dataset (Ekman's six emotions plus *anticipation, optimism, pessimism, love,* and *trust*), leading to the creation of 11 separate datasets. This approach allowed for formulating the emotions identification as a binary classification task. Ezen-Can and Can focused on emojis, hashtags, and textual content. They divided the dataset into training, validation, and test subsets. Recognising the significance of visual elements in conveying emotions, the study paid particular attention to emojis, uncovering several interesting patterns across different emotion categories. Trust was rarely indicated through emojis, whereas *sadness* prevalently used the sobbing face emoji and, surprisingly, the laughing with joy emoji. The weary face emoji also appeared frequently in this category. *Anger* and *disgust* predominantly featured the laughing with joy emoji, suggesting possible irony, followed by the sobbing face emoji. The rage emoji is the third most used. Love had the heart eyes emoji, but this emoji was also common in joyful tweets. *Joy* and *optimism* had a significant presence of the fire emoji. Joy was also the emotion with the highest number of emojis. Ezen-Can and Can developed separate RNNs models for each emotion category. During testing, the predictions from each model were combined by tweet. The strength of the methodology lies in the autonomy of each binary classification model, permitting diverse architectures and parameters tailored to each emotion category. These models comprised three GRU layers, including two bidirectional layers, each containing 100 neurons, and the final encoding layer had 50 neurons. The model achieved a 0.398 accuracy in emotion detection and a F1 score of 0.358. This performance notably surpassed a random baseline, which stopped at 0.185 accuracy and 0.285 F1. Furthermore, the results demonstrated that specific emotion categories, such as surprise and trust -covering only about 4–5% of all the training dataset-presented a significant challenge due to the scarcity of examples. Such imbalance led the classifiers to generate more false negatives. However, one advantage of using binary models in this context is their ability to handle the disparate distribution of examples across emotion and emotion categories more adeptly.

Rathnayaka et al. (2019) introduced the Pyramid Attention Network (PAN) for detecting emotions in microblogs. This method's primary benefit is the model's ability to analyse sentences from various angles, enabling the identification of multiple emotions. The first attention layer focuses on word embeddings and the initial GRU layer, while the second attention layer added a second GRU layer. The PAN model was evaluated using the SemEval 2018 emotion classification dataset (Mohammad et al., 2018), which comprises 10,983 tweets, each potentially expressing one or more out of eleven emotions. Given the possibility of multiple emotions per tweet, Rathnayaka et al. applied multi-label classification methods. The dataset was divided into training, development, and testing sets, containing 6,838, 886, and 3,259 tweets respectively. Rathnayaka et al. benchmarked the model's performance against the top four models from SemEval 2018. The results demonstrated that the PAN model achieved state-of-the-art performance on the emotion detection of the SemEval-2018's dataset.

Polignano et al. (2019) developed a deep neural network integrating Bi-LSTM, CNN, and a self-attention layer, and tested its efficacy across various emotion datasets. They also examined the performance of three distinct pre-trained word embeddings for encoding words. The initial layer of the model converted input sentences into word embeddings vectors, suitable for neural network processing. Utilising pre-trained word embeddings from varied domains. They compared three pre-trained word embeddings: Google word embeddings, GloVe, and FastText. For sentence transformation into word embeddings, sentences were tokenised using NLTK's TweetTokenizer. Analysing the model's outcomes across three datasets, Polignano et al. found that the word embedding vector with the best results was FastText in all the three evaluated datasets: ISEAR, SemEval 2018 and SemEval 2019.

In the list below, we summarise the methodologies and findings of the studies we just reviewed.

Emotion detection from TV show transcripts—Shrivastava et al., 2019:

Model/Approach: Sequential CNN with attention mechanism
Findings/Contributions: CNN model outperformed LSTM and RF classifiers in emotion detection.

Emotion detection in tweets—Ezen-Can and Can, 2018:

Model/Approach: Series of binary classifiers for each of 11 emotion categories
Findings/Contributions: Demonstrated effectiveness of binary classification for each emotion category, with specific focus on visual elements like emojis.

Emotion detection in human-chatbot interactions—Basile et al., 2019:

Model/Approach: Four neural network models including bidirectional LSTM with attention
Findings/Contributions: Explored different configurations for emotion detection, achieving best performance with a feed-forward network and BERT.

Evaluation of word embeddings models for emotion detection—Polignano et al., 2019:

Model/Approach: Deep neural network integrating Bi-LSTM, CNN, and self-attention layer
Findings/Contributions: FastText was the most effective word embedding model for the detection of emotions across three datasets.

Emotion detection in microblogs—Rathnayaka et al., 2019:

Model/Approach: Pyramid Attention Network (PAN)
Findings/Contributions: PAN model capable of detecting multiple emotions in texts, achieving state-of-the-art performance on the SemEval 2018 dataset.

Emotion detection in Weibo microblogs—Liu et al. (2021):

Model/Approach: Semantic and Emoticon Based Emotion Recognizer (SEER) with Bi-GRU and attention

Findings/Contributions: the SEER model combines Bi-GRU network and emoticon analysis for emotion detection, outperforming the other tested methods.

6.1.2   Deep Learning for the Detection of Emotion Dimensions

In the realm of emotion dimension detection, deep learning models categorise text according to emotional valence (positive or negative sentiment), arousal (intensity of emotion), and dominance (control or power, or lack thereof, expressed by the emotion). By analysing the linguistic patterns and contextual clues within the text, these models can effectively discern subtle variations in emotional intensity, such as distinguishing between *joy* and *elation* or *sadness* and *despair*. Furthermore, integrating attention mechanisms in the deep learning models has enhanced their ability to focus on relevant parts of the text, thereby improving the accuracy of emotion detection. These mechanisms allow the models to weigh different parts of the input text differently, giving more importance to words or phrases that indicate emotional content.

Wang et al. (2016) introduced a regional CNN-LSTM model aimed at predicting the Valence-Arousal ratings of texts. Differing from the traditional CNN that processes the entire text as a single input, Wang et al. treated each sentence as a separate region with their CNN model (regional CNN). First, Wang et al. built word vectors for the vocabulary using word embedding techniques. The regional CNN then used these vectors to create text vectors for the texts under prediction. In contrast to the traditional CNN approach that processes the entire text in one go, the regional CNN treated each sentence as a distinct region. This method broke down the text into several regions, allowing for extracting and emphasising the affective information from the regions. In the final step, the regional information was sequentially integrated using LSTM for Valence-Arousal space prediction. The integration of the regional CNN with LSTM allowed Wang et al. to take into consideration both the intra-sentence regional details and the inter-sentence long-distance dependencies within the prediction process. The performance of the regional CNN-LSTM model was evaluated by comparing it with various lexicon-based and neural network techniques for predicting valence-arousal in both English and Chinese datasets. Wang et al.'s regional method demonstrated superior performance compared to other neural network models, highlighting the efficiency of sequentially integrating regional information. Additionally, the Pearson correlation coefficient for arousal prediction was lower than for valence prediction, suggesting that predicting arousal in texts poses more challenges.

Lakomkin et al. (2017) devised a method for determining the intensity of emotions in tweet messages as a part of the WASSA 2017 EmoInt shared task. Participants had text of tweets and their classified emotions (e.g., *anger, joy, fear*, or *sadness*). Their task was to develop a system capable of assigning values to the intensity of these emotions. Lakomkin et al. proposed an ensemble combining two neural network models. This ensemble operates at the character (e.g., punctuation and emoticons) and word levels and integrates a lexicon approach. In the specific context of determining the emotion intensity in tweets, a character-level model has the potential to effectively recognise unique aspects of social media communication, such as hashtags, emoticons, or repeated characters. This model is adept at detecting prevalent writing styles in social media, especially punctuation and other characters.

Conversely, a word-level recurrent neural model leverages the sequential arrangement of text, utilising distributed word representations trained on extensive textual data. The described method leverages a dual-component ensemble to predict the intensity of emotions in tweets. For the character-based model, Lakomkin et al. used an LSTM network. This model is specifically trained on a vast collection of tweets, enabling it to process and understand text at the character level. The word-level model employed a bidirectional GRU network and word embedding. This model processes the tweets by considering the context and order of words, thereby offering a deeper understanding of the emotional content conveyed by the text.

These two models, each focusing on different levels of text processing, are integrated into an ensemble. The ensemble approach combines the outputs of the character-based and word-level models to yield a more nuanced and accurate prediction of emotional intensity in tweets. The model evaluation was based on the correlation rank coefficients, which measure the relevance and similarity between two sets of rankings. Although the neural models achieved lower correlation values than the baseline model, which was rich in external knowledge, the ensemble's performance was particularly notable in the subset of data with moderate to high emotional intensities. Here, it showed an approximate 18.5% relative improvement in both correlation coefficients. Interestingly, the character-level model yielded competitive and, in some cases, superior results for fear and joy emotions, especially in high-intensity samples.

Goel and colleagues (2017) developed an ensemble integrating three distinct methodologies in the same competition. The methods involved three different approaches to analysing emotions in tweets:

1. Feed-forward Neural Network: this approach used a neural network architecture to predict emotions from tweets. A 443-dimensional vector represented each tweet, combining Word2Vec representations with the TweetToLexiconFeatureVector, which computes features from several lexicons. The network architecture consists of four hidden layers with Rectified Linear Unit (ReLU) activation and dropout to avoid overfitting.
2. Multitask Deep Learning: this method employed multitask learning with deep neural networks. The network's initial layers were shared across multiple emotions to increase generalisation, while the top layers are emotion specific. The same features as in Approach 1 were used, and the network was trained through four cycles for each emotion. This approach aimed to learn generalised features across different emotions while focusing on emotion-specific aspects.

3. Sequence Modelling using CNNs and LSTMs: this approach utilised a combination of CNNs and LSTM neural networks. Word2Vec embeddings were used to represent words in tweets, and these embeddings were concatenated to form the input. The network architecture varied between LSTMs, CNNs, or a combination of both, followed by fully connected layers. The network was trained to minimise the error between the actual and predicted values of emotion intensity, using a similar training process as in the first approach.

The ensemble model developed outperformed the baseline model significantly, achieving average scores of 75.26% and 74.70% in cross-validation and test sets, respectively. This improvement, approximately 14% and 10% higher than the baseline, underscores the superiority of deep learning models over lexicon-based approaches. The CNNs and LSTM networks performed the best of the three methods used. Regarding the performance across different emotions, the ensemble model was most effective in detecting the intensity for the emotions of *sadness*, followed closely by *fear*, then *joy*, and finally *anger*.

In the context of the same competition, Zhang et al. (2017) presented a CNN model with LSTM. The CNN extracted local n-gram features within tweets, and the LSTM captured the word dependency across the tweets. The model effectively captured local and long-range dependencies within the tweets for emotion intensity, particularly *anger* and *joy*.

Wu et al. (2019) introduced a semi-supervised method for dimensional sentiment analysis using a variational autoencoder. Usually, variational autoencoders learn the distribution of the input data and, sampling from the distribution, produce new instances. However, Wu et al. used the variational autoencoder in a different fashion. Their model had three modules: encoding, valence-arousal prediction, and decoding. The encoding module employed LSTM to transform the input texts into vectors. Wu et al. used a 2-layer Bi-LSTM to estimate sentiment scores across different dimensions for valence-arousal prediction. The decoding module used these outputs to reconstruct the original texts using LSTM. A crucial aspect of the method was the reconstruction of both labelled and unlabelled texts by the decoding module.

Consequently, this approach leveraged the vast pool of unlabelled data to enhance the training of the dimensional sentiment model. Tests were conducted on three benchmark datasets: the Facebook post dataset (Preoţiuc-Pietro et al., 2016), the Chinese Valence and Arousal dataset (Yu et al., 2016), and the EmoBank dataset (Buechel & Hahn, 2017). The results showed that Wu et al.'s method significantly boosted emotion dimensions prediction, even in cases when training data were scarce, with respect to many other neural network methods, including Wang et al.'s (2016).

Atmaja et al. (2019) focused on recognising categorical and dimensional emotions in both Text-to-Speech spoken and written texts. Atmaja et al. developed a deep-learning text emotion recognition system with an RNN and GRU. This method was tested on three text datasets containing sentences labelled with categorical and dimensional emotions: ISEAR, IEMOCAP (Busso et al., 2008), and EmoBank. A variety of metrics were employed to test the model, including accuracy. The findings reveal that texts containing indirect emotional information, such as spoken or expressively written experiences, are more easily recognised than straightforward texts, such as news headlines. In tasks involving categorical and dimensional emotion recognition, the system performed better with fewer categories than with a larger number. The study also pointed out that, besides the dataset size, the number of categories within the dataset impacted the performance of categorical emotion detection. A smaller dataset with fewer categories yielded better results than a larger, category-rich, but imbalanced dataset. In contrast, the larger datasets had a lower error rate for dimensional emotion detection.

In the following, we summarise the models, methodologies, and findings of the articles reviewed in this section:

Valence-Arousal Prediction—Wang et al., 2016:

Model/Approach: Regional CNN-LSTM, treats each sentence as a separate region.
Findings/Contributions: Superior to other models, especially in valence prediction; difficulty in arousal prediction.

Ensemble of Neural Networks—Lakomkin et al., 2017:

Model/Approach: Combining character-level (LSTM) and word-level (Bi-GRU) models; it includes a lexicon approach.
Findings/Contributions: 18.5% improvement in high-intensity emotions detection; effective for *fear* and *joy*.

Ensemble with Various Approaches—Goel et al., 2017:

Model/Approach: Feed-forward NN, Multitask Deep Learning, CNNs & LSTMs.
Findings/Contributions: Significant improvement over baseline; best for *sadness, fear, joy*, and *anger detection.*

CNN with LSTM—Zhang et al., 2017:

Model/Approach: Extracts local n-gram features; it captures word dependencies.
Findings/Contributions: Effective in *anger* and *joy* intensity detection.

Semi-Supervised VAE—Wu et al., 2019:

Model/Approach: Encoding; Valence-Arousal prediction, Decoding with LSTM and Bi-LSTM.

Findings/Contributions: Enhanced emotion dimensions prediction, effective with limited data as well.

RNN with GRU—Atmaja et al., 2019:

Model/Approach: Tested on spoken and written text; it works better with fewer categories.
Findings/Contributions: Better recognition of indirect emotional information; larger datasets favoured in dimensional tasks.

6.2   Advantages and Limitations of Machine Learning Approaches

Deep learning models have undeniably revolutionized emotion detection in NLP, and they have significant advantages and notable limitations. One of the primary strengths of deep learning is the ability to recognize complex patterns in text data, making them highly effective at identifying subtle emotional nuances. Additionally, these models are adept at contextual understanding, enabling them to interpret emotions based on the surrounding text. This contextualization is particularly valuable in understanding the nuances of human emotions. Moreover, deep learning approaches excel in handling large datasets and *Big Data*. Deep learning models learn from extensive and diverse text sources, which enhances their accuracy and reliability. They are also adaptable to different languages, provided they are trained with sufficient data, making them versatile in multi-lingual tasks. Furthermore, as more data becomes available, deep learning models can continually learn and adapt, improving their performance over time.

However, deep learning models face numerous challenges as well. They are heavily dependent on the quality and quantity of the training data, and any biases present in the data can lead to inaccurate detection in the test datasets. Additionally, these models require significant computational resources, which may not be accessible to all researchers or organizations. Another critical issue is their lack of explainability; often operating as *black boxes*, it can be challenging to discern how these models arrive at specific conclusions, a significant drawback for sensitive applications (e.g., detection of emotions in mental health applications). There is also the risk of overfitting the models; that is, the models might perform exceptionally well on training data but fail to generalize to new datasets. We outline the key strengths and weaknesses of deep learning in the context of emotion detection in the following:

Accuracy

Advantages: High accuracy and handling of complexity.
Limitations: Risk of overfitting; can struggle with subtlety and nuance.

Context Analysis

Advantages: Good at understanding context and relationships in text.
Limitations: Challenges in interpreting cultural nuances (e.g., *sarcasm* and *irony*).

Scalability

Advantages: Efficient in handling large datasets.
Limitations: Requires significant computational resources.

Language Adaptability

Advantages: Capable of adapting to different languages and slang.
Limitations: Limited effectiveness in less resourced languages.

Learning Capability

Advantages: Continuously improves with more data.
Limitations: Dependent on the quality and quantity of training data.

Transparency

Limitations: Lack of explainability in the decision-making processes.

Bias

Limitations: Potential for bias, based on the training data.

To overcome some of these limitations, transformers and generative models have been introduced. Thanks to the training on extensive datasets, which can encompass a wide range of linguistic, cultural, and emotional contexts, transformer models such as BERT (Bidirectional Encoder Representations from Transformers) and RoBERTa (Robustly Optimized BERT Pretraining Approach), and generative models such as GPT (Generative Pretrained Transformer) can offer improved contextual understanding and the ability to accurately capture nuanced emotional expressions.

6.3 Transformers and Generative Models

Emotion detection in NLP has witnessed a remarkable evolution in recent years. As the nuances of human emotions demand that models comprehend not just the text's literal meaning but also the context and cultural nuances, transfer learning can partially overcome these limitations using pre-trained models. Transfer learning uses large datasets from different tasks and applies the learned knowledge to related but distinct problems. This approach allows the leverage of previously trained models to enhance performance with reduced reliance on labelled data. It involves transferring knowledge from a domain with ample data samples, numerous labels, and high-quality annotation to a related but distinct target domain with limited data samples, fewer labels, or noisy data. Consequently, transfer learning facilitates the detection, understanding and interpretation of emotion subtleties without the need for extensive new training data and resources.

In transformer models (Vaswani et al., 2023), self-attention allows each word in an input sequence to be evaluated in the context of every other word. The self-attention mechanism generates a set of attention scores that are then transformed into a weighted sum, reflecting how much focus to place on each part of the data when interpreting a specific element. Thus, the output combines information across the sequence, enabling the transformer model to understand and represent the context effectively. Devlin et al. (2019) introduced BERT (Bidirectional Encoder Representations from Transformers), a revolutionary method for language understanding. BERT processes text in a deeply bidirectional manner, evaluating both preceding and subsequent context in text interpretation. BERT is trained on vast datasets, and the training involves two phases: pre-training on unlabelled data and fine-tuning for specific NLP applications. Liu et al. (2019) advanced the BERT methodology with RoBERTa, incorporating a broader dataset and refined parameters. RoBERTa's training leveraged extensive English datasets, including the English Wikipedia, BooksCorpus, and additional web-sourced datasets. Another incarnation of BERT, DistilBERT (Sanh et al., 2020), aims to reduce BERT's size while preserving its effectiveness. It simplifies BERT's structure by halving the layers and omitting some of the components, resulting in a faster, smaller model suitable for general applications. Like BERT, DistilBERT is trained on English Wikipedia and BooksCorpus datasets. XLNet (Yang et al., 2020), another variant, acquires contexts bidirectionally using different permutations of the factorisation order.

Yang et al. (2019) introduced EmotionX-KU, a contextual emotion classifier, utilising the EmotionLines dataset (Chen et al., 2018), which comprises two subsets: the Friends and EmotionPush subsets. EmotionLines categorises emotions into four labels: *neutral, joy, sadness*, and *anger*. A notable characteristic of the dataset is the uneven distribution of these emotion categories, with 'neutral' being the most prevalent, followed by *joy, sadness*, and *anger*. In their approach, Yang et al. (2019) treated the task of contextual emotion classification as a sequence problem, considering each utterance within the context of a dialogue. Their model employed transfer language modelling and dynamic max pooling, accounting for both the utterances and their contextual information. The authors conducted experiments by training their model on both subsets, together and separately, to compare the outcomes. The authors hypothesised that the uncased BERT-base model, a self-attention-based transferable language model, would yield more accurate predictions of contextual emotions after post-training. However, they noted a limitation in their model: it could not process all utterances when the number of input tokens surpassed the maximum length set in the model. This constraint highlighted a potential area for future enhancement in handling larger input sequences.

Huang et al. (2019) detected emotion in the EmotionLines dataset using a three-stage model: the casual utterances phase, pre-training, and fine-tuning. The first phase focused on casual utterances, aimed to retain emotional content within dialogues by amalgamating a casual utterance and its reply into a single combined utterance. The experiments demonstrated that this type of sentence mapping for continuous dialogue enabled the BERT model to effectively capture the emotions in the reply sentence. The method achieved F1 scores of 0.815 on the Friends subset and 0.885 on the EmotionPush subset.

Adoma et al. (2020) evaluated the performance of BERT, RoBERTa, DistilBERT, and XLNet in detecting emotions using the ISEAR dataset and aiming to categorise emotions into seven categories: *anger, disgust, sadness, fear, joy, shame*, and *guilt*. Employing identical hyperparameters, the models achieved varying degrees of accuracy. In descending order, the degrees of accuracy were the following: RoBERTa (accuracy = 0.7431), XLNet (0.7299), BERT (0.7009), and DistilBERT (0.6693).

As regards the detection of emotion dimensions, Park et al. (2021) developed a framework aimed at learning dimensional Valence-Arousal-Dominance (VAD) scores from annotated datasets labelled with categorical emotion tags. Their approach involved using a fine-tuned, pre-trained BERT model to achieve two main objectives: categorical classification of emotions and simultaneous prediction of dimensional VAD scores. They utilised the NRC-VAD lexicon for mapping categorical labels to VAD dimensions, enabling the learning of the VAD distributions from the categorical labels of the input sentences. The training of the model involved categorising emotions in the VAD space based on their corresponding scores. Park et al. applied their model to various datasets, including SemEval 2018, ISEAR, and EmoBank. SemEval 2018 comprises 10,983 tweets labelled with either *neutral* or *no emotion* or one of 11 other emotions. The ISEAR dataset contained labels for seven emotions: *joy, sadness, fear, anger, guilt, disgust*, and *shame*, totalling 7,665 sentences. EmoBank includes over 10,000 sentences dimensionally annotated as per the VAD model, with a subset of categorically annotated emotions that used Ekman's basic emotion model, making it ideal for dual representational analysis. Moreover, Park et al. added a linear transformation layer to the BERT model. They initially trained it on the ISEAR and part of the SemEval data, then fine-tuned it using the SemEval dataset. The model's performance was evaluated using the EmoBank data, achieving F1 scores of 0.688 and 0.695 when fine-tuned on ISEAR and SemEval, respectively. When trained on SemEval and fine-tuned on EmoBank, the model yielded F1 scores of 0.659, 0.327, and 0.287, outperforming the state-of-the-art VAD regression models.

The Generative Pre-trained Transformers (GPT), designed by OpenAI (Radford et al., 2018; Radford et al., 2019), is an attention-based model devoided of recurrent layers. It combines learned word embedding positions with token embeddings, feeding them into a transformer decoder for language modelling. GPT operates as a unidirectional attention model, using transformer decoders in a semi-supervised learning approach. The GPT-3 model (Brown et al., 2020) represents an expansion of GPT-2, featuring a staggering 175 billion trainable parameters. Its architecture mirrors GPT-2, with the notable distinction of the transformer layers, which adopt a pattern of alternating dense and locally banded sparse attention. GPT-3's training encompassed eight versions of varying sizes, ranging from 125 million to 175 billion parameters. One of the key advantages of the GPTs is the enhanced lexical robustness and the flexibility of the pre-trained model to adapt to various tasks without needing specific model alterations. Notably, the GPT models performed well over several domain-specific models, achieving state-of-the-art results in a wide array of language modelling tasks across different domains.

Elyoseph et al. (2023) tested ChatGPT's capability in recognising and articulating emotions. Elyoseph et al.'s study employed the emotional awareness scale to assess ChatGPT's responses to twenty different scenarios objectively. The aim was to compare its emotional awareness performance against general population norms established by prior research. Additionally, a follow-up test was conducted after one month to evaluate any improvements in emotional awareness over time. Two independent licensed psychologists also assessed the relevance and appropriateness of ChatGPT's emotional awareness responses. The initial assessment showed that ChatGPT's emotional awareness scores were significantly higher than the general population's (Z score = 2.84). In the subsequent test, ChatGPT's performance was further enhanced, nearly reaching the highest possible emotional awareness score (Z score = 4.26), with exceptionally high accuracy levels (9.7/10). The findings indicate that ChatGPT can produce fitting emotional awareness responses and that its proficiency in this area can improve significantly over time. This research has theoretical and practical implications, suggesting that ChatGPT support psychiatric diagnoses and assessments and enriches emotional language.

Wang et al. (2023) tested whether GPT 3.5 could analyse emotions in memes. Memes combine textual and visual elements, and they are a potent yet intricate medium for conveying thoughts and emotions and require an understanding of social and cultural nuances. A particular challenge in this domain is identifying and moderating hateful memes, which often carry offensive content in subtle forms. Wang et al. (2023) examined GPT-3.5's effectiveness in such nuanced tasks, including meme sentiment classification, humour type identification, and implicit hate detection in memes. By comparing the model's performance with human annotations of the datasets from SemEval-2020 and the Facebook's hateful memes dataset, Wang et al. gained insight into GPT 3.5's capabilities and limitations. Despite GPT 3.5's impressive achievements, the model struggled with subjective tasks due to inherent limitations like contextual comprehension, interpretation of underlying meanings, and data biases.

Kheiri and Karimi (2023) extensively analysed different GPT models for sentiment analysis, focusing on the SemEval 2017 dataset. The study explores three approaches: 1) prompt of GPT-3.5 Turbo, 2) fine-tuning of various GPT models, and 3) implementing word embedding classification. The study benchmarked these GPT-based techniques against other high-performing models previously applied to the same dataset. The findings demonstrated a marked superiority of GPT methods in predictive accuracy, achieving over 22% higher F1-score than the current state-of-the-art. Additionally, Kheiri and Karimi discussed the challenges in contextual understanding and sarcasm detection, highlighting the GPT models' advanced ability to navigate these issues with respect to previous models.

However, GPT models are not without drawbacks. One significant limitation is their resource-intensive nature, particularly during the pre-training phase, which can be costly. Additionally, these models face challenges in modelling dependencies that exceed predetermined fixed lengths. They are also computationally demanding and costly not only to train but also to maintain.

In the following, we provide a concise overview of each study, the methods used, the key findings, and their limitations.

BERT—EmotionX-KU—Yang et al. (2019):

Method: Contextual emotion classifier using the EmotionLines dataset; considers utterances within dialogue context.
Findings: Effective in contextual emotion classification.
Study Limitations: Limitation in handling larger input sequences beyond a set maximum length.

BERT—Huang et al. (2019):

Method: Three-stage model for emotion detection in EmotionLines dataset.
Findings: Focuses effectively on emotional detection between two consecutive utterances (utterance and reply).
Study Limitations: Limited scope in terms of only focusing on two consecutive utterances, may not capture the full context of longer dialogues.

BERT, RoBERTa, XLNet—Adoma et al. (2020):

Method: Evaluated various models on ISEAR dataset for emotion detection; varying accuracies with RoBERTa showing highest performance.
Findings: Demonstrated strong performance in emotion detection, especially with RoBERTa.

Study Limitations: Limited to seven specific emotion categories, may not encompass the full spectrum of human emotions.

BERT—Park et al. (2021):

Method: Uses BERT for learning Valence-Arousal-Dominance (VAD) scores from categorical emotion tags; trained on various datasets.
Findings: Effective in learning complex emotion scores from text.
Study Limitations: Requires explicit VAD annotations for optimal performance; generalizability may be limited to datasets similar to those used in training.

GPT-3—Brown et al., (2020):

Method: 175 billion parameters, alternating dense and locally banded sparse attention patterns in transformer layers.
Findings: Advanced capabilities in handling large-scale language modeling.
Study Limitations: Resource-intensive nature; struggles with modeling dependencies longer than designated fixed lengths.

GPT-3.5—Elyoseph et al., (2023):

Method: Assessed emotional awareness capabilities using emotional awareness scale; significant improvement over time.
Findings: Showcased significant improvements in emotional awareness.
Study Limitations: High accuracy levels might not translate directly to real-world scenarios.

GPT-3.5—Wang et al. (2023):

Method: Investigated GPT-3.5's effectiveness in tasks such as meme sentiment classification, humour type identification, and implicit hate detection.
Findings: Demonstrated capability in diverse and complex language tasks.
Study Limitations: Struggles with subjective tasks due to inherent limitations such as interpretation of underlying meanings.

GPT-3.5 and 3—Kheiri and Karimi (2023):

Method: Analysed different GPT methods in sentiment analysis; benchmarked against other models on SemEval 2017 dataset.
Findings: Showed strengths in sentiment analysis across diverse contexts.
Study Limitations: Challenges in contextual understanding and sarcasm detection.

A recent review by Brin et al. (2023) indicated that GPT-4 demonstrates elements of cognitive empathy, such as emotion recognition and providing emotionally supportive responses, particularly in healthcare contexts. In some cases, GPT-4 outperformed medical doctors in empathy-related tasks, exhibiting cognitive empathy across various health and mental health scenarios.

Given the promising capabilities of ChatGPT-3.5 and GPT-4 in emotion detection and cognitive empathy, significant ethical concerns arise. Privacy and data security are critical, as handling sensitive data about the user's emotions requires stringent measures to protect the user's information. There is also a risk of over-reliance on AI for emotional support, which can diminish human judgement's autonomy. Additionally, emotional data could be misused by malicious actors, underscoring the need for ethical frameworks and regulatory oversight to ensure responsible use of these technologies. Finally, the possible data bias in AI systems must be addressed to prevent perpetuating societal biases and ensure equal treatment for all users.

In the next and final Chapter, we will focus on the concept of explainable AI and the ethics of AI in emotion detection, discussing the challenges and strategies for creating emotion detection AIs that comply with ethical guidelines, safeguard user's well-being, and maintain public trust.

References

Adoma, A. F., Henrym, N.-M., & Chen, W. (2020). Comparative analyses of Bert, Roberta, Distilbert, and Xlnet for text-based emotion recognition. *17th International Computer Conference on Wavelet Active Media Technology and Information Processing*, 117–121.

Atmaja, T., B., Kiyoaki, S., & Masato, A. (2019). Speech emotion recognition using speech feature and word embedding. In *2019 Asia-Pacific Signal and Information Processing Association Annual Summit and Conference* (pp. 519–523).

Basile, A., Franco-Salvador, M., Pawar, N., Štajner, S., Chinea Rios, M., & Benajiba, Y. (2019). SymantoResearch at SemEval-2019 task 3: Combined neural models for emotion classification in human-chatbot conversations. *Proceedings of the 13th International Workshop on Semantic Evaluation*, 330–334.

Brin, D., Sorin, V., Vaid, A., Soroush, A., Glicksberg, B. S., Charney, A., W., Nadkarni, G., & Klang, E. (2023). Comparing ChatGPT and GPT-4 performance in USMLE soft skill assessments. *Scientific Reports, 13*(1), 1–5.

Brown, T. B., Mann, B., Ryder, N., Subbiah, M., Kaplan, J., et al. (2020). Language models are few-shot learners. *ArXiv, 2005*, 14165.

Busso, C., Bulut, M., Lee, C. C., et al. (2008). IEMOCAP: Interactive emotional dyadic motion capture database. *Language Resources and Evaluation, 42*, 335–359.

Buechel, S., & Hahn, U. (2017). EmoBank: Studying the Impact of Annotation Perspective and Representation Format on Dimensional Emotion Analysis. In *European Chapter of the Association of Computational Linguistics '17*. 10.18653/v1/E17-2092.

Cer, D., Yang, Y., Kong, S., Hua, N., et al. (2018). Universal sentence encoder for English. *Proceedings of the 2018 Conference on Empirical Methods in Natural Language Processing, System Demonstrations*, 169–174.

Chen, S.Y., Hsu, C-C., Kuo, C-C., Ku, L-W. et al. (2018). *Emotionlines: An emotion corpus of multi-party conversations.* arXiv:1802.08379, 2018.

Cho, K., van Merrienboer, B., Bahdanau, D., & Bengio, Y. (2014). *On the properties of neural machine translation: Encoder-decoder approaches.* arXiv:1409.1259.

Chung, J., Gulcehre, C., Cho, K., & Bengio, Y. (2014). Empirical evaluation of gated recurrent neural networks on sequence modeling. *NIPS 2014 Workshop on Deep Learning.*

Devlin, J., Chang, M-W., Lee, K., & Toutanova, K. (2019). BERT: Pre-training of deep bidirectional transformers for language understanding. *Proceedings of the 2019 Conference of the North American Chapter of the Association for Computational Linguistics: Human Language Technologies,* Vol.1, 4171–4186.

Elyoseph, Z., Hadar-Shoval, D., Asraf, K., & Lvovsky, M. (2023). ChatGPT outperforms humans in emotional awareness evaluations. *Frontiers in Psychology, 14,* 1199058.

Ezen-Can, A., & Can, E. F. (2018). RNN for affects at SemEval-2018 task 1: Formulating affect identification as a binary classification problem. *Proceedings of the 12th International Workshop on Semantic Evaluation,* 162–166.

Farzi, R., & Bolandi, V. (2016). Estimation of organic facies using ensemble methods in comparison with conventional intelligent approaches: A case study of the South Pars Gas Field, Persian Gulf, Iran. *Modeling Earth Systems and Environment, 2,* 105.

Ghosh, L., Saha, S., & Konar, A. (2020). Bi-directional long short-term memory model to analyze psychological effects on gamers. *Applied Soft Computing, 95.*

Goel, P, Kulshreshtha, D., Jain, P., & Shukla, K. K. (2017). Prayas at EmoInt 2017: An ensemble of deep neural architectures for emotion intensity prediction in tweets. *Proceedings of the 8th Workshop on Computational Approaches to Subjectivity, Sentiment and Social Media Analysis,* 58–65.

Huang, Y-H., Lee, S.-R., Ma, M-Y., Chen, Y-S., Yu, Y-W., & Chen, Y-S. (2019). *EmotionX-IDEA: Emotion BERT – an affectional model for conversation.* arXiv:1908.06264v1.

Kheiri, K., & Karimi, H. (2023). *Exploiting GPT for advanced sentiment analysis and its departure from current machine.* arXiv 2307.10234.

Lakomkin, E., Bothe, C., & Wermter, S. (2017). GradAscent at EmoInt-2017: Character and word level recurrent neural network models for tweet emotion intensity detection. *Proceedings of the 8th Workshop on Computational Approaches to Subjectivity, Sentiment and Social Media Analysis,* 169–174.

Li, W., Qi, F., Tang, M., & Yu, Z. (2020). Bidirectional LSTM with self-attention mechanism and multi-channel features for sentiment classification. *Neurocomputing, 387,* 63–77.

Liu Y., Ott M., Goyal N., Du J., Joshi M., Chen D., Levy O., Lewis M., Zettlemoyer L., & Stoyanov V. (2019). *Roberta: A robustly optimized BERT pretraining approach.* arXiv:1907.11692

Liu, C., Liu, T., Yang, S., & Du, Y. (2021). Individual Emotion Recognition Approach Combined Gated Recurrent Unit With Emoticon Distribution Model. *IEEE Access,* 1–1. 10.1109/ACCESS.2021.3124585.

Machová, K., Szabóova, M., Paralič, J., & Mičko, J. (2023). Detection of emotion by text analysis using machine learning. *Frontiers in Psychology, 14,* 1190326.

Mohammad, S., Bravo-Marquez, F., Salameh, M., & Kiritchenko, S. (2018). SemEval-2018 task 1: Affect in tweets. In *Proceedings of the 12th International Workshop on Semantic Evaluation* (pp. 1–17).

Park, S., Kim, J., Ye, S., Jeon, J., Park, H-Y., & Oh, A. (2021). Dimensional emotion detection from categorical emotion. In *Proceedings of the 2021 Conference on Empirical Methods in Natural Language Processing* (pp. 4367–4380).

Preoţiuc-Pietro, D., Schwartz, H. A., Park, G., Eichstaedt, J. Kern, M., Ungar, L., & Shulman, E. (2016). Modelling Valence and Arousal in Facebook posts. In Proceedings of the 7th Workshop on Computational Approaches to Subjectivity, Sentiment and Social Media Analysis.

Polignano, M., de Gemmis, M., & Semeraro, G. (2019). SWAP at SemEval-2019 task 3: Emotion detection in conversations through tweets, CNN and LSTM deep neural networks. In *Proceedings of the 13th International Workshop on Semantic Evaluation* (pp. 324–329).

Radford, A., Narasimhan, K., Salimans, T., & Sutskever, I. (2018). *Improving Language Understanding by Generative Pre-Training.* https://gwern.net/doc/www/s3-us-west-2.amazonaws.com/d73fdc5ffa8627bce44dcda2fc012da638ffb158.pdf

Radford, A., Wu, J., Child, R., Luan, D., Amodei, D., & Sutskever, I. (2019). *Language models are unsupervised multitask learners.* https://d4mucfpksywv.cloudfront.net/better-language-models/language-models.pdf

Rathnayaka, P., Abeysinghe, S., Samarajeewa, C., Manchanayake, I., Walpola, M. J., Nawaratne, R., Bandaragoda, T. R., & Alahakoon, D. (2019). *Gated recurrent neural network approach for multilabel emotion detection in microblogs.* arXiv:1907.07653

Sanh, V., Debut, L., Chaumond, J., & Wolf, T. (2020). *DistilBERT, a distilled version of BERT: smaller, faster, cheaper, and lighter.* arXiv 1910.01108

Shrivastava, K., Kumar, S., & Jain, D. K. (2019). An effective approach for emotion detection in multimedia text data using sequence based convolutional neural network. *Multimedia Tools and Applications, 78*(20), 29607–29639.

Vaswani, A., Shazeer, N., Parmar, N., Uszkoreit, J., & Jones, L., et al. (2023). *Attention is all you need.* arXiv 1706.03762

Wang, J., Yu, L-C., Lai, K. R., & Zhang, X. (2016). Dimensional sentiment analysis using a regional CNN-LSTM model. *Proceedings of the 54th Annual Meeting of the Association for Computational Linguistics,* Vol.2, 225–230.

Wang, J., Luo, J., Yang, G., Hong, A., & Luo, F. (2023). *Is GPT powerful enough to analyze the emotions of memes?* arXiv 2311.00223

Wu, C., Wu, F., Wu, S., Yuan, Z., Liu, J., & Huang, Y. (2019). Semi-supervised dimensional sentiment analysis with variational autoencoder. *Knowledge-Based Systems, 165*, 30–39.

Yang, K., Dongyub L., Taesun W., Seolhwa L., & Heuiseok L. (2019). *EmotionX-KU: BERT-Max based contextual emotion classifier.* arXiv.

Yang, Z., Dai, Z., Yang, Y., Carbonell, J., Salakhutdinov, R., & Le, Q. L. (2020). *XLNet: Generalized autoregressive pretraining for language understanding.* arXiv 1906.08237

Yu, L-C., Lee, L-H., Hao, S., Wang, J., He, Y., Hu, J., Lai, K. & Zhang, X. (2016). Building Chinese Affective Resources in Valence-Arousal Dimensions. In Proceedings of NAACL-HLT'16. 10.18653/v1/N16-1066.

Zhang, Y., Yuan, H., Wang, J., & Zhang, X. (2017). YNU-HPCC at EmoInt-2017: Using a CNN-LSTM model for sentiment intensity prediction. In *Proceedings of the 8th Workshop on Computational Approaches to Subjectivity, Sentiment and Social Media Analysis* (pp. 200–204).

Challenges and Ethical Considerations of Emotion Detection

7

7.1 Explainable Artificial Intelligence

As we have seen in Chap. 6, despite advancements in model accuracy, deep learning and machine learning models lack the necessary human interpretability and transparency. Machine learning models are often black boxes due to their non-linear and nested structures, which obscure the understanding of how information flows from inputs such as text to predicted outputs.

The concept of explainable AI addresses this issue, advocating for machine learning and deep learning models to be transparent about their decision-making processes and actions. As we explore explainable AI, we will examine its critical role in ensuring that AI systems, and specifically AI systems for emotion detection, operate within ethical boundaries. The issue of AI transparency includes addressing the challenges of actually creating transparent models, the importance of multidisciplinary collaborations for explainable AI, and the compliance with regulatory standards, particularly in sensitive domains such as healthcare. We will then discuss the ethical challenges of developing NLP applications for emotion detection, focusing on data biases, the risk of misuse, and the necessity for ethically responsible AI development. By integrating ethical principles throughout the AI development lifecycle, we can promote fairness, accountability, and transparency in AI systems, ultimately safeguarding the user's well-being and maintaining public trust.

Research supports this approach; for instance, Zucco et al. (2018) highlighted the necessity of explainable sentiment analysis in medical applications to reliably extract knowledge about opinions and emotions from user-generated text, ensuring transparency and trustworthiness. Similarly, Turcan et al. (2021) examined the use of more explainable models for detecting psychological stress in online posts, demonstrating how clear and understandable AI models can enhance ethical practices by providing transparency in decision-making processes. Yang et al. (2019) presented a pioneering model designed to

F. Cavicchio, *Emotion Detection in Natural Language Processing*, Synthesis Lectures on Human Language Technologies, https://doi.org/10.1007/978-3-031-72047-5_7

predict and categorise emotions based on their intensities and link them to the topics that evoke these emotions. Their approach utilised an innovative interpretable neural network with the unique feature of initialising its hidden layer in a manner akin to topic models, a concept drawn from transfer learning. Additionally, they crafted a specialised error function to optimise the network's focus on effectively ranking emotions related to specific text content. This model is dubbed Interpretable Relevant Emotion Ranking with Event-Driven Attention (IRER-EA). IRER-EA's architecture includes four key layers. The Input Embedding Layer integrates word and event embeddings. Following this, the Encoder Layer features two encoders, one dedicated to words and one to events. The Attention Layer, a pivotal third layer, calculates attention scores at both word and event levels, ensuring the model accounts for relevant information across both corpus and document scales. The process culminates in the Output Layer, which synthesises the insights from earlier stages to produce the final emotion rankings. This layered approach empowers the IRER-EA model to rank emotions effectively while focusing on texts' most relevant event-driven content. Consequently, the model excelled in identifying topics and words related to emotions, reflecting real-life events, and aligning with human intuition.

The IRER-EA model was tested on three diverse real-world datasets. The first dataset, Sina Social News (Zhou et al., 2018), comprises 5,586 news articles from Sina's Society channel, where readers voted on six emotions: *funny, moved, angry, sad, puzzled*, and *shocked*. The second dataset, the Ren-CECps corpus (Quan & Ren, 2010), is a Chinese collection of 1,487 blogs, each annotated with eight basic emotions (*anger, anxiety, hope, hate, joy, love, sorrow,* and *surprise*) from the writer's perspective, with a scoring system to indicate the intensity of these emotions. The third dataset used was SemEval 07 (Strapparava & Mihalcea, 2007), further diversifying the model's testing grounds.

Gadeka and Guelorget (2020) developed an approach to deep learning classification focusing on the writing style of texts to identify emotions and detect fake news with explainable AI. Their method utilised a Textual Class-activation-mappings Convolutional Neural Network (TC-CNN) designed for effectively categorising texts by recognising patterns in groups of consecutive words. To extend its capabilities, they introduced a comprehensive framework to analyse texts for emotional tone, bias, and indications of hatred. This framework was then applied in a case study to examine the writing styles in recent news articles, including those flagged as fake news, across major English-language news outlets. As addressed by Gadeka and Guelorget, the challenge in explainable AI classification lies in offering detailed explanations for why texts were categorised into specific classes, which is more complex than basic labelling. They explored two main methods for this analysis. The first method examined the role of an attention mechanism in text classification, where high attention scores were assigned to keywords. While helpful, attention scores alone did not provide specific explanations for each classification label, only indicating the general importance of parts of the document.

The second method employed Class Activation Maps (CAMs), which are more specific in determining how various document segments contribute to its classification. For their analysis, Gadeka and Guelorget used two datasets: Kaggle's fake news dataset with 13,000 English documents labelled as fake news and the Signal-Media dataset, a collection of legitimate press articles. These datasets were used to analyse the prevalence and influence of emotional content within the articles. They then tested the model on various news outlets, from microblogging websites such as Gab Trends to classic news outlets such as the BBC. Their findings indicated a predominance of *sadness* in the emotional spectrum of the datasets, unexpectedly overshadowing the *neutral* category. This trend probably stems from the generally serious themes of politics, geopolitics, and economics in the news. The analysis maintained the original categorisation of emotions, with *hate* indicating *anger* and a separate hate label marking the presence of hateful or offensive content. Interestingly, there was a notable occurrence of either *happy* or *hateful* articles in the Gab Trends data subset. Notably, The Onion, known for its satirical content, showed *sadness* levels comparable to serious news outlets, although the model did not detect humour. Regarding bias and factual reporting, the model identified classic newspapers such as the Nypost, Daily Mail, and Fox as factual despite their varied journalistic styles. Major networks such as CNN and BBC were also categorised as predominantly factual. The data analysis revealed that traditional factual articles typically evoke *sadness*, whereas biased articles tend to incite *happiness* or *anger*. In the following we presents an overview of the studies reviewed, specifically outlining each study's topic, objectives, and key findings.

Sentiment Analysis—Zucco et al. (2018):

Objective: Explainable sentiment analysis in medical applications.
Findings: Reliable extraction of opinions and emotions from user-generated text.

Emotion Intensity Prediction and Categorization—Yang et al. (2019):

Objective: Innovative interpretable neural network for emotion prediction and categorization.
Findings: Identification of topics and words related to emotion intensity.

Biased News Identification through Emotions—Gadeka and Guelorget (2020):

Objective: Focus on writing style for emotion identification and bias identification.
Findings: High occurrence of *happy* and *angry* for biased articles.

Stress Detection in Online Posts—Turcan et al. (2021):

Objective: Using explainable models for detecting psychological stress.
Findings: Comparable results to state-of-the-art BERT. The model infused with emotion words mirrors psychological components of stress.

7.2   Ethical Aspects of Emotion Detection

As we have seen in Chap. 6, AI and machine learning are increasingly employed to under-stand and predict human emotions, with applications expanding in mental healthcare, education, online recruitment, and automotive design. In healthcare, emotion detection aims to personalise treatment and potentially enhancing patient health outcomes. However, concerns arise from the limitations of these technologies, such as a narrow emotion range and potential patient discomfort (Pavez et al., 2023). Health professionals have expressed apprehensions about the validity and reliability of emotion detection via serious games, raising issues regarding the range of emotions detected and the potential discomfort it may cause patients.

Additionally, McStay (2019) and Steinert and Friedrich (2020) highlighted the risk of reinforcing stereotypes, particularly concerning vulnerable groups like children, neu-rodivergent individuals, and senior citizens (Bryant & Howard, 2019; Ienca & Malgieri, 2022). These concerns emphasise the need for rigorous ethical standards and the inclusion of diverse perspectives in developing and deploying these technologies.

Emotion detection technologies often promise benefits like improved user experience and safety, yet their underlying assumptions need to be revised. As demonstrated by Facebook's 2014 *emotional contagion study*, ethical implications are sometimes over-looked, which raised significant ethical concerns due to its manipulative nature and lack of informed consent (Kramer et al., 2014). Moreover, while social media responses can be measured, they do not always reflect actual emotional states, making the interpretation of such data challenging (Kohavi et al., 2020). The sensitivity of emotion-related data presents further ethical concerns, as emotions can infer mental states closely linked to human identity and autonomy. The public and clients often view the use of emotional data as potentially manipulative, raising concerns about privacy and consent (Andalibi & Buss, 2020).

Another significant challenge in emotion detection studies is the need for demographic diversity in the annotated datasets used to train algorithms. These datasets often draw from limited sources like movies, TV shows, and social media posts, failing to account for the vast differences in emotional expression across various demographic groups, such as those defined by age, gender, ethnicity, and cultural background. As a result, algorithms trained on these datasets can exhibit biases and inaccuracies. For instance, Buolamwini and Gebru (2018) found higher error rates for facial recognition algorithms when iden-tifying individuals with darker skin tones. Similarly, an AI recruitment tool by Amazon was discontinued due to bias against women, penalising resumes with expressions such as *women's colleges* (Dastin, 2018). In healthcare, Obermeyer et al. (2019) discovered biases in algorithms used to allocate healthcare resources, which were less likely to refer black patients than white patients with the same health status because they used health costs as a proxy for health needs, overlooking disparities in access to healthcare.

To identify biases in sentiment analysis, Kiritchenko and Mohammad (2018) created the Equity Evaluation Corpus (EEC). By generating 8,640 sentences with gender- and race-associated words, the authors evaluated 219 systems and found significant biases. These systems often rated sentiments differently based on the gender or race mentioned in the sentences, highlighting the need for tools like the EEC to detect and address these biases. The lack of diverse datasets can lead to algorithms that fail to generalise well across different populations, perpetuating societal biases and leading to unfair outcomes.

Bias in the emotion detection models is another critical issue. For example, Ekman and Friesen's (1971) model, though validated across cultures for facial and vocal recognition (Sauter et al., 2010), is criticised for its Western-centric approach and lack of cultural context consideration (Ong, 2021). Facial expressions of emotions can be modified to conform to societal and cultural norms, further complicating the accuracy of emotion detection systems (Matsumoto & Hwang, 2013). Furthermore, Ekman's theory of universal basic emotions does not account for emotions such as *Schadenfreude* (the *glee felt for someone else's misfortune* in German) and *amae* (the *feeling of being taken care of completely and unconditionally* in Japanese) are rarely included in standard emotion datasets, leading to a limited understanding of human emotions. The lack of such diverse emotional concepts limits current emotion detection models, thus perpetuating a limited and potentially biased understanding of human emotions in NLP applications.

The reliance on Ekman's universal model has led to a partial neglect of the variability in emotional expression across different demographic groups. Studies show that emotion annotation can vary significantly based on the demographic profile of the annotators (Plaza-del-Arco et al., 2024). This subjectivity underscores the importance of considering who produces and annotates the data we use to train the algorithms. Furthermore, the predominance of negative emotions in the datasets can skew the performance of the algorithms, limiting their effectiveness in various contexts. For example, detecting fine-grained positive emotions might be more relevant in educational settings where positive reinforcement can help the learning process.

Mohammad (2022) introduced Ethics Sheets for automatic emotion recognition in AI tasks to promote ethical design and implementation of AI systems. This framework addresses various ethical factors throughout the AI development process, including the justification for automating a task, the impact on different groups of people, and the potential for misuse. Key questions addressed by the Ethics Sheets include the inherent ambiguity and unpredictability of human behaviour related to the task, the theoretical foundations, the social and cultural influences on task design, data, methodology, and evaluation, the potential abuses of automated systems. The Ethics Sheets consider theoretical foundations and ethical implications, including human variability and privacy, and the AI system results' explainability and visualisation.

Building on the research presented, future research should prioritise the collection and integration of demographic information from both data creators and annotators, promoting inclusivity and reducing bias. Expanding the range of emotion categories to include those from diverse psychological theories would provide a more comprehensive understanding of emotions. Interdisciplinary research incorporating insights from psychology, cognitive science, and philosophy could further enrich emotion detection models in NLP. By addressing these gaps, the field of emotion detection in NLP can advance towards creating more effective and inclusive tools for understanding human emotions.

References

Andalibi, N., & Buss, J. (2020). The human in emotion recognition on social media: Attitudes, outcomes, risks. In *Proceedings of the ACM SIGCHI Conference on Human Factors in Computing Systems.*

Bryant, D., & Howard, A. (2019). A comparative analysis of emotion-detecting AI systems with respect to algorithm performance and dataset diversity. In *Proceedings of the 2019 AAAI/ACM Conference on AI, Ethics, and Society* (pp. 377–382).

Buolamwini, J., & Gebru, T. (2018). Gender shades: Intersectional accuracy disparities in commercial gender classification. *Proceedings of Machine Learning Research, 81*, 1–15.

Dastin, J. (2018). *Amazon scraps secret AI recruiting tool that showed bias against women.* Reuters, October 10, 2018. https://www.reuters.com/article/us-amazon-com-jobs-automation-insight-idU SKCN1MK08G

Ekman, P., & Friesen, W. V. (1971). Constants across cultures in the face and emotion. *Journal of Personality and Social Psychology, 17*(2), 124–129.

Gadek, G., & Guelorget, P. (2020). An interpretable model to measure fakeness and emotion in news. *Procedia Computer Science, 176*, 78–87.

Ienca, M., & Malgieri, G. (2022). Mental data protection and the GDPR. *Journal of Law Bioscience, 9*(1), lsac006.

Kiritchenko, S., & Mohammad, S. (2018). Examining gender and race bias in two hundred sentiment analysis systems. In *Proceedings of the Seventh Joint Conference on Lexical and Computational Semantics* (pp. 43–53).

Kohavi, R., Tang, D., & Xu, Y. (2020). *Trustworthy online controlled experiments: A practical guide to A/B testing.* Cambridge University Press.

Kramer, A., Guillory, J., & Hancock, J. (2014). Experimental evidence of massive-scale emotional contagion through social networks. *Proceedings of the National Academy of Sciences, 111*(24), 8788–8790.

Matsumoto, D., & Hwang, H. S. (2013). Cultural influences on nonverbal behavior. In D. Matsumoto, M. G. Frank, & H. S. Hwang (Eds.), *Nonverbal communication: Science and applications* (pp. 97–120). Sage Publications.

McStay, A. (2019). Emotional AI and EdTech: serving the public good? *Learning, Media and Technology, 45*(3), 270–283. https://doi.org/10.1080/17439884.2020.1686016

McStay, A., & Rosner, G. (2021). Emotional artificial intelligence in children's toys and devices: Ethics, governance and practical remedies. *Big Data & Society.*

Mohammad, S. M. (2022). Ethics sheet for automatic emotion recognition and sentiment analysis. *Computational Linguistics, 48*(2), 239–278.

Obermeyer, Z., Powers, B., Vogeli, C., & Mullainathan, S. (2019). Dissecting racial bias in an algorithm used to manage the health of populations. *Science, 366*(6464), 447–453.

Öhman, E. (2020). Emotion annotation: Rethinking emotion categorization. In *DHN Post-Proceedings* (pp. 134–144).

Ong, D. C. (2021). An ethical framework for guiding the development of affectively-aware artificial intelligence. In *9th International Conference on Affective Computing and Intelligent Interaction.* IEEE.

Pavez, R., Diaz, J., Arango-Lopez, J., Ahumada, D., MendezSandoval, C., & Moreira, F. (2023). Emo-mirror: A proposal to support emotion recognition in children with autism spectrum disorders. *Neural Computing and Application, 35*(11), 7913–7924.

Plaza-del-Arco, F.M, Cercas Curry, A. A., Cercas Curry, A., & Hovy, D. (2024). Emotion analysis in NLP: Trends, gaps and roadmap for future directions. In *Proceedings of the 2024 Joint International Conference on Computational Linguistics, Language Resources and Evaluation* (pp. 5696–5710).

Quan, C., & Ren, F. (2010). Automatic annotation of word emotion in sentences based on Ren-CECps. In *Proceedings of the Seventh International Conference on Language Resources and Evaluation.*

Sauter, D. A., Eisner, F., Ekman, P., & Scott, S. K. (2010). Cross-cultural recognition of basic emotions through nonverbal emotional vocalizations. *Proceedings of the National Academy of Science, 107*(6), 2408–2412.

Steinert, S., & Friedrich, O. (2020). Wired emotions: Ethical issues of affective brain-computer interfaces. *Science and Engineering Ethics, 26*, 351–367.

Strapparava, C. & Mihalcea, R. (2007). SemEval-2007 task 14: Affective text. In *Proceedings of the Fourth International Workshop on Semantic Evaluations* (pp. 70–74).

Turcan, E., Muresan, S., & McKeown, K. (2021). Emotion-infused models for explainable psychological stress detection. In *Proceedings of the 2021 Conference of the North American Chapter of the Association for Computational Linguistics: Human Language Technologies* (pp. 2895–2909).

Yang, Y. Zhou, D., He, S., & Zhang, M. (2019). Interpretable relevant emotion ranking with event-driven attention. In *Proceedings of the 2019 Conference on Empirical Methods in Natural Language Processing and the 9th International and Joint Conference on Natural Language Processing* (pp. 177–187).

Zhou, D., Yang, Y., & He, Y. (2018). Relevant emotion ranking from text constrained with emotion relationships. In *Meeting of the North American Chapter of the Association for Computation Linguistics* (pp. 561–571).

Zucco, C., Liang, H., Di Fatta, G., & Cannataro, M. (2018). Explainable sentiment analysis with applications in medicine. In *IEEE International Conference on Bioinformatics and Biomedicine* (pp. 1740–1747).